TURNED UPSIDE DOWN

# TURNED UPSIDE DOWN

## Marko Pogačnik

*Translated by Tony Mitton*

A Workbook on Earth Changes
and Personal Transformation

LINDISFARNE BOOKS

Copyright © 2004 by Marko Pogačnik
Translation copyright © 2004 by Lindisfarne Books
First published in German as *Erdwandlung als persönliche Herausforderung* by Droemer Knaur, 2003

Published by Lindisfarne Books
400 Main Street
Gt. Barrington, MA 01230
www.lindisfarne.org

ISBN 1-58420-025-1

Library of Congress Cataloging-in-Publication Data available

All rights reserved. No part of this book may be reproduced in any form without the written permission of the publisher, except for brief quotations embodied in critical reviews and articles.

10 9 8 7 6 5 4 3 2 1

Printed in the United States of America by
The P.A. Hutchison Company

100% post-consumer waste paper

Dedicated to all the men and women, known and unknown, who are our co-workers worldwide, and, whether as individuals meditating in the Silence or gathered in groups for earth healing and geomancy, are opening themselves to the true Earth and, in dialogue with the rich realms of their lives, are feeling their way forward.

# Table of Contents

**Introduction** .................................................. 1

**1. It's Time to Change: Renew Your Life!** ...................... 3
*The Message from the Inner Child* ............................... 3
*The Human Holon* ............................................... 23
*The Elemental and Spiritual Aspects of Identity* ................ 30
*Self-Recognition in the Mirror of Earth Changes* ................ 35
*Helpers Stand Ready* ........................................... 39

**2. The Fundamental Causes Leading to the Catastrophe
of September 11** ............................................... 45
*The Unexpected Gift of Catharsis* ............................... 45
*The Powers That Work Against the Earth Changes* ................ 59

**3. After the Catastrophe: Humanity's
New Inner Organization** ........................................ 69
*The Cosmic Double* ............................................. 69
*The Missing Energy Channels* ................................... 78
*The Heart Is the Midpoint* ..................................... 89

**4. February 2002: The Decisive Somersault in Etheric Space** ... 99
*The Genetic Code of the Currently Approaching Somersault* ...... 99
*How the Reversal of Etheric Space Invaded Everyday Life* ....... 106
*A Renewed Earth Cosmos* ........................................ 120
*The Healing Centers* ........................................... 124

**5. After the Somersault: The Rushing Torrent
of the Great Purification** .................................129
*The Double's Shadow Aspect* ..................................129
*The Cunning of the Contrary Powers* ..........................143
*A Clear Ethical Stance*.......................................153
*Clear Emotions* ..............................................154

**6. The Consciousness of Earth Organizes
Anew Its Plane of Feeling** ...................................157
*Elemental Beings in the Service of Charitable Love* ..................157
*Special Changes in the Consciousness of Earth* .....................172
*The Transformed Ego of Humankind* ............................183

**7. The Next Goal of the Processes of Change** .................193
*The Arrival of the Redemptress*................................193
*Preparations for the Next Phase of Change* ......................205
*The Future Scene of Change: The Spiritual-Emotional Plane* .........209

**8. The Present State of Affairs** ............................219
*Problems with the Exaggerated Yang* ............................219
*One More Warning*...........................................223
*Release from One's Own Entanglements with the Dark Powers* .......227
*Centering Earth's Etheric Space* ................................233

**Appendix: Overview of the Exercises** ........................237
1. The Inner Child ............................................237
2. The Human Holon ..........................................242
3. The Various Extensions of the Human Being ..................248
4. The Heart Is the Midpoint ..................................252
5. The Somersault of Space ....................................254
6. Clarification and Transformation of Shadow Aspects ..........255
7. The Seven Foundation Stones of the New Ethic ..............262
8. The Changed Elemental Beings ..............................263
9. Discovery of the Spiritual-Emotional Plane ..................265

**Bibliography** ..............................................271

## INTRODUCTION

THIS BOOK IS CONCERNED WITH the fact, completely overlooked by most of humanity, that for the past several years our planet has been undergoing an intensive inner transformation. Up till now the transformation process appears to have had no serious effect on our everyday life. This is deceptive, because the future of our common earthly living space will be completely redefined by the Earth changes that are even now drawing to completion. *The Earth will no longer be as it once was, even a few years ago.*

I have the good fortune to practice an unusual profession, one in which I constantly attune to the different planes and dimensions of earth space and observe their qualities. My business is with landscape healing, lithopuncture and related geomantic research in various sites, towns and countries. I have now been doing this artistically stimulating work for more than two decades. In the course of it I discovered that during 1997 the Earth, acting as a common consciousness, initiated a self-healing process that avoided the anticipated danger of worldwide destruction brought about by human activity—good tidings for us but hard for the mind to comprehend.

My book is built up of several layers. The foundation layer is composed from about forty dreams through which I obtained insights into the still hidden processes of change in Earth and humanity. Threaded within that is the layer of my perceptions and experiences that ground the messages in my dreams. Added to these are about fifty drawings through which I wish to include the nonverbal plane of communication. In several chapters I propose exercises to give the reader an opportunity to personally experience and incorporate particular themes. The Appendix brings these together and adds further exercises that will enable readers to immerse themselves holistically in the experiences recounted in the book.

The present book follows on my three earlier books dealing with the first four years of the astonishing processes of Earth change. The book entitled *Christ Power and the Earth Goddess* describes the first months of the process (1997–1998); *Earth Changes, Human Destiny* is concerned with the years 1998–2000; *Daughter of Gaia* focuses, among other things, on the changes that came to completion in the year 2000. This new book deals with the period from the spring of 2001 to the spring of 2003—a tense time during which the Earth introduced some quite new and unexpected permutations into the change process—all of which present a complex personal challenge.

There is one more point. This book will be difficult to grasp if you read it exclusively with your mind. I am an artist, and my way of communicating is primarily through images that will unfold themselves completely in the accompanying stream of feelings and imagination. At the same time I am a *male* and strive to provide the mind—that masculine-imprinted left half of the brain—with many bits of information.

To help readers enjoy the book, I recommend that you give yourselves over to the stream of narrative. The mind can be reassured with the promise that, on further reading, it can ponder the details.

Sempas, Slovenia
December 9, 2002

Marko Pogačnik

CHAPTER ONE

# It's Time to Change: Renew your Life!

### The Message from the Inner Child

EVER SINCE EARTH BEGAN TO make its inner changes, I have seen the repetition of a certain pattern in my life: No sooner have I got clear about the personal changes necessary to adapt to the Earth's current phase and adjusted myself to the complicated threads of its process than I am thrust into renewed uncertainty. Signs, dreams and events emerge to make me feel that the Earth's development has taken a new direction and my hard-won sense of security is once again in jeopardy.

So I ask, what direction is it taking now? Which plane of being will it touch this time? What can one do to tune into its new route? One would certainly like to avoid the threatened renewal of personal inner conflicts and disruptions, but how?

This was my pattern that kept repeating during the spring of 2001. I was then deeply engaged in a rich stream of creative pursuits when an impressive dream burst right into their midst. I had just sent off the manuscript of my book *Daughter of Gaia*, in which I present the current process of Earth change as a self-initiated planetary rebirth. It was natural for me to feel well versed in the theme, but it was precisely this sense of security that the dream put in question.

The longer and more deeply I thought over the dream's images and their accompanying feelings, and cudgeled my head over their possible meanings, the less they corresponded to the archetypes of personal change with which I had been familiar up till then. I noted in my diary that it was obvious that we were about to uncover a dimension of the human Self of which we hitherto had had scarcely any conception.

In the dream, which I had when I was in Lucerne on May 16, 2001, I first see a great number of naked men marching in a long column, one behind the other. These are not young men; in fact they are very old. Behind each man a small child is marching, just as naked as he. Suddenly, a man breaks up this ordered rhythm; I notice that his belly is unusually swollen, as if he were pregnant. Quickly turning around, he lifts up the child marching at his back and breaks out of the line. He has obviously decided to go his own way, and he takes off with the child in his arms.

I am watching this scene with my beloved wife at my side. Our attention is drawn to the actions of this unknown man and we quickly discuss whether to intervene, given the possibility that the man is stealing a child that does not belong to him. However, we decide to act as if we have noticed nothing unusual.

This dream was deeply significant to me. A few years earlier, when I was planning my book *Christ Power and the Earth Goddess*, I had followed up every lead I found on the theme of the "inner child." The symbol of the child in its various connections is emphasized in all the Gospels, to say nothing of the multiple representations of the Mother of God in those parts of the world permeated by Catholicism. Whichever way we turn, the Madonna, so to speak, has our inner child tied to her apron strings.

The theme of the dream comes very close to the words of Jesus quoted in the Gospel according to Thomas: "The man old in days will not hesitate to ask a small child seven days old about the place of life, and he will live. For many who are first will become last, and they will become one and the same."

Fundamental elements of this saying make their appearance in

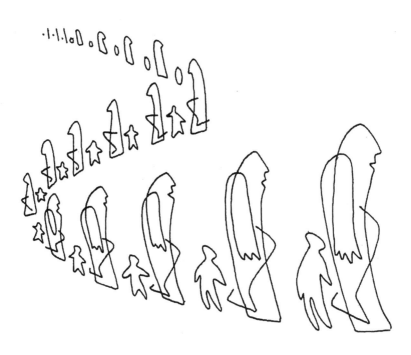

*The dream image shows many elderly men marching forward in a column, and at the back of each there walks a child.*

my dream. I am thinking in particular of the deep emotion that was inspired by the old man's union with the small child. Although my role in the dream was that of a remote observer, I could not but notice the portentous quality of the meeting that was taking place before my eyes. "The small child seven days old" who was lifted up and taken into the arms of the strange man represents the dimension of eternity within the human being. Before we enter into the processes of incarnation and birth, we have an unlimited share in eternity and are quite obviously integrated with the polyphony of the cosmic music. As soon as we are born as a human being, our space turns a somersault and that which was within goes outside. The freedom of the universal spirit is packed into the confines of an incarnation. From now on we must adjust ourselves to a relatively narrow structure composed of time and space. The newborn human soul is bound space-wise into the coordinates of matter and time-wise into the restricting rhythm of the clock.

In contrast with the restrictions of incarnation, "the small child seven days old" represents an abundance of cosmic connectedness that is still consciously experienced. This child carries within itself the whole knowledge of our true origins and life's meaning, and also the power to transplant that knowledge, free of the limitations of space and time, into the human being. Our upbringing in childhood and the resultant "extorted adaptation" to cultural and social norms and patterns cause us to lose the unerring infallibility of "the small child seven days old"; that is to say, we lose the capacity to mirror back the wholeness of the cosmos.

The emphasis of Jesus' words is to ask us to recognize that it is an illusion for our mind to believe that the cosmic quality of the child within us has been lost through the process of upbringing and maturation. Indeed, that is never lost. It is the divine core of our being and still beats at our midpoint. From the standpoint of Mind, however, this quality has been banished to a dimension that the incarnated human being finds nearly unapproachable. This now banished extension of our being is also called the Higher Self.

*A fourteenth-century Greek icon provides an example of space turning a somersault: In this image the fontanel of the Jesus child is being brought into relationship with the soles of his feet.*

Jesus' words assure us that the eternal aspect of humankind is directly within our reach. Whether we experience ourselves as being adult or even as elderly, we human beings have the gift in every moment of enjoying the timeless wisdom of our Higher Self. Jesus, Master of the Western culture, also informs us as to the means by which we can become one with eternity while still remaining within the structure of time. He points to the principle of the somersault, of being turned upside down: "For many who are first will become last." Here we are obviously dealing with a nonlogical principle of thought and action that can lead us across into the Universal Oneness: "And they will become one and the same."

But how can one apply this principle to one's daily practice?

In my dream that somersault, so full of promise, is represented symbolically by one's back. To reunite with the quality of eternity, one must turn round and reach into one's posterior space. That is where we will find the inner child who makes possible the return of our consciousness to the source of Being.

Here, the posterior space stands for the unconscious—in contrast to our body's anterior space, which is easily controlled by Mind and is symbolized by the logic of the waking consciousness.

Put into everyday language, Jesus' words should encourage us to release ourselves for a moment from the light of the intellect and dive into the dark unconscious, there to touch eternity and bring it into Time. The "First," i.e., the light of the intellect, will thus become the "Last," which is symbolically represented by the unconscious. Thus, it is through this backwards somersault that we will finally become one with the Universal Wholeness.

Quite emphatically, this does not involve a simple return to the realm of the unconscious. Rather, it involves a somersault in the truest sense of the word, by which the "First" not only becomes "Last" as described above, but is also turned upside down. We gradually learn to incorporate the quality of eternity within the everyday light of logical thinking.

Jesus' words in regard to the inner child imply the basis for a

"Western yoga," and encourage one to take one's own individual spiritual path. However, this involves a serious problem, which in the dream vision is represented by me.

In the dream, instead of participating and rejoicing over the reunion, I stand to one side and observe the goings-on with a very critical eye. I even worry explicitly that a criminal act is in process. I am almost ready to call in the police. Yet I have to admit that in the "pregnant-looking" man I recognize my own naked form, just as I see it in the bathroom mirror. That I was looking at myself, even though I did not realize it, is also the reason why I did not sound the alarm but opted to ignore the event.

In fact I succeeded in ignoring the dream for more than another month, though I certainly recorded the observations in my diary and began to contemplate them philosophically. I ignored it because I felt I did not know what might be my task in connection with it. The momentous contrast between my two roles in the dream and the inner conflict that resulted did not enter my consciousness.

I became conscious of it in an unexpected manner on July 20 of the same year. On that day I was in Slovenia with my daughter and coworker Ajra Miska, holding a seminar in our study course "Education in Geomancy and Personal Development." Ajra had already developed various exercises for me on the theme of the inner child, and one of these was presented for the experience of our group. In brief, the exercise runs as follows: One imagines oneself to be standing at a window in a dark room, watching oneself as a small child playing in the garden outside. At the moment that feels instinctively right, one decides to open the door, enter the garden and go to the child. Then, one should observe precisely how the child reacts.

When I began to meet my inner child in this exercise, I had the sensation of being several meters taller. Worse, the nearer I drew to the child, the taller I became. The child seemed to be lost far underneath me. Then, with its tender hands, the child touched my endlessly long legs. To my horror, they were stiff and wooden.

I was shattered by the shocking disproportion between my personality, ruled by the dry mind, and the inner child, finely attuned to the realms of feeling. In that moment I decided to turn my life completely around—with the goal of building a dam around the power of my mental consciousness and giving space to the voice of eternity within me.

The decision appeared to fall on fruitful ground, for three days later I had an experience that took me nearer to my goal.

The manner in which the experience came to me is quite telling: I was en route from Ljubljana to Reykjavik, where I would lead a geomantically inspired trek in Iceland. To be at the airport on time, I was to stay overnight in Ljubljana with my daughter Nike's family. I had been told I would be welcome but that unfortunately no bed was available, only my six-year-old grandson's narrow cot.

Curled up in the child's bed, I slept dreamlessly through the first half of the night. Then, in the middle of the night I woke up and felt something warm and gentle in my belly region. I stretched myself out and could sense the inner child lying on its back in the area of my belly. It had adopted the pre-birth position, the top of its head resting against my pubic bone. It was at once clear to me that this experience was a gift from the little boy whose bed I was borrowing for the night.

I lay there quietly in order to get a sense of the sort of dimension that was opening up within me. Suddenly, there began the process of inversion. The child made something like a backwards somersault—all within the watery realm of my lower body—turned itself around like a fish in water and stretched out, landing in the middle of my upper body. Now I could feel the top of its head near my throat and its root chakra in the region of my solar plexus. The water had vanished. The presence of the inner child complemented my feeling that my inner self had had a wake-up call and was now focused in my heart-center.

More than a year later I led another seminar group to Medjugorje. This lies in the country called Herzegovina, which, together with Bosnia, forms one of the new states—like my homeland Slovenia—that have come into being since the collapse of Yugoslavia. The place has become celebrated throughout the world, ever since the Virgin Mother of God appeared fifteen years ago to a group of children on a nearby hill. This conversation between the visible and invisible dimensions has been continued through today. Relative to our theme of the inner child, it may be remarked that the children through whom this communication was begun have in the meantime grown up to have families and the usual, everyday kinds of responsibilities.

It is significant that it was only at her first appearance in Medjugorje that the Universal Mother showed herself with her divine child. Clearly, she wished the human children of Herzegovina to identify her with the image of the Mother of God known to them in church. She has since regularly appeared unaccompanied, sending each one of us a clear message that the time has come to take over responsibility for the Self, permeated by our spiritualized soul. No longer will she be our proxy, holding us all in her lap.

This revolutionary decision on the part of the divine Virgin is pictured on the right-hand side of the altarpiece in the Basilica at Medjugorje. Clad in her blue cloak, Mary floats in the middle of the composition, and both her hands are free to pray for the fate of humanity. Her hands are free because a man, St. Anthony, is standing nearby, and is filled with happiness to be allowed to hold her child in his arms.

The picture inspired me to take the seminar participants to the church in Medjugorje and propose the following exercise:

- You sit for a few minutes sunk deep in inner silence. Then, when you are ready to hold the divine child in your lap, gently stretch out your arms and ask the Mother of Wholeness, the divine Virgin, to give the child to you.
- You hold the child for a while so as to feel the cosmic quality that streams from within the Christ Child.

*A detail from the altar in the Basilica at Medjugorje: The Virgin Mary has given her child to St. Anthony, so as to have both her hands free and be able to pray for world peace.*

- Suddenly, the child in your lap appears to see something interesting on the floor and bends down to touch it—finally bending so far forward that the top of its head points to the Earth's center. Thus it happens that the child turns upside down, as suggested by Jesus' words: What was highest will be lowest, thereby thrusting open the doorway to eternity.
- At this moment you should envision that the child's body turns quickly around as if making a somersault and rises up through your own bodily structure. This motion is the exact opposite of the birth process. You are, so to speak, newborn, but this time not through your physical mother's birth canal but by consciously turning round and taking your own path.
- The child is now in the middle of your inner space. You feel the challenge to let go of material forms and projections and instead concentrate on experiencing the qualities of emotions, soul and spirit. How does it feel for the divine core to be awakened within you? It gives you free access to the original space of eternity. What are you going to bring there with you?

Two months later I was shown a deeper aspect of this exercise. While I was in Prague preparing for my second seminar on town healing, I discovered a shrine of the Black Madonna not far from Hradcany, as the Prague castle is called. Hidden in the midst of the former convent of Maria Loreto, a feminine counterpoint to the powerful castle, is a chapel with a sculpture of the Black Madonna and the risen Child. This image shows the right hand of the Madonna lightly touching the sole of the divine Child's foot. This is taking place in the intimate region of the lower belly, to very discreetly indicate the inversion whereby the higher becomes lower. One has the impression that the Madonna's finger is touching specific acupressure points on the sole of the foot of the inner child, in order to consciously connect herself with certain powers and qualities of the divine.

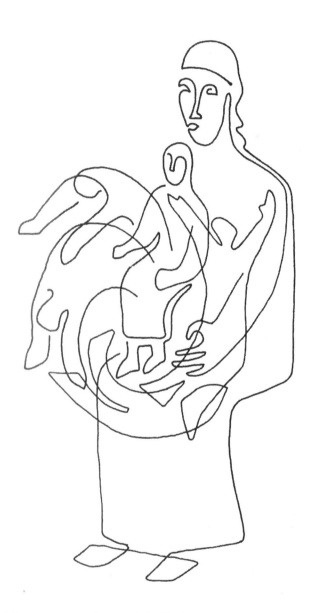

*A Romanesque sculpture of the Madonna from southern France was the model for the exercise in which the inner child turns upside down.*

This portrayal inspired me to extend and deepen the exercise described above:

- Imagine that the inner child is fully present and standing in your midsection.
- Now imagine yourself to be touching various points on the soles of the child's foot with the sensitive fingertips of both your hands. It is recognized that specific zones on the soles of the feet resonate with various physical organs and functions. Relating this to the divine Child, it means that specific points on the soles of its feet resonate with the "organs" and "functions" of the Universal Wholeness, or, more precisely, with the various powers that ensoul the universe.
- Through the soles of the inner child's feet, connect and anchor yourself to the multidimensionality of life.

It should be mentioned that Prague does contain a shrine dedicated to the inner child. According to tradition the little sculpture of the boy Jesus, the so-called Infant Jesus of Prague, came there as a gift from Spain, and over past centuries has been the object of passionate devotion. It is to be found in the Church of Our Lady of Victory, Santa Maria de Victoria. There are not many other churches that hold a solitary image of the Jesus child—with neither Joseph nor Mary at his side—standing there for himself alone in his own autonomous space and full dignity.

It should now be apparent that the Jesus child symbolizes the perfection and original strength of our true Self.

I had planned to take the seminar group into this church and there, in the quiet of the shrine, practice the exercise of inverting the inner child. However, scarcely had each of us found our seats and started to settle into the exercise than a group of English-speaking pilgrims began a thunderous prayer, amplified by a loudspeaker. Here two worlds were really exploding against each other! There was our group trying to experience the reality of the inner child, while the other was making an outer racket to pray to the same divine principle. It is true that both paths should lead to the same goal, but their discrepancy was nearly intolerable.

Oppressed by the noise of the loudspeaker, the inner child—this is how I experienced the event—lay down in one half of a seedlike capsule and covered itself with the other half.

I began to look around me with the eyes of the soul. What I perceived was very encouraging: In every one of those present—including the tourists wandering around the church unaware—I saw just such a seed vibrate, with the inner child slumbering in its capsule. It was patiently waiting to be discovered and raised to awareness.

According to this, the message of my dream about the "pregnant" man tells us that a time is coming—or for many has already come—when the inner child is no longer content to slumber in one's inner room. It wants to wake up and integrate itself with the individual's life processes. Moreover, the message continues, one will not be able to cope with the challenges of the continuing Earth changes if one's own inner self is not awake.

Unhappily, in modern society the quest for the inner child is too often perverted to pedophilia. Those affected, instead of working to demolish the blockages separating them from the blessed sense of union with their own inner child, project outwards their need to experience wholeness. Grown men who "love" little children have mistaken the path. The inner qualities for which they yearn are choked in the external and material. Pedophilia is a shadow accompanying the all too common alienation of modern humans, whose relationship to their innermost core is almost extinguished.

Each one of us contributes to this shadow when we shy away from our own inner truth—it is not only those who abuse children sexually or in other ways that bear some guilt. And is it not also true of us when we pray to "the small child seven days old" outside of us, but do not at the same time look for it within us?

If we are to continue our search for the inner child, it is absolutely necessary for us to be clear about the danger of making our own contribution to the shadow of pedophilia. I had a dream related to this subject on December 9, 2002, when I was busily engaged in writing the present chapter.

The dream image was about the process of inverting the inner child as described above. The somersault-like inversion took place so extremely slowly that there was no way I could overlook the child's gender. This was not a boy but a girl.

The archetypal image of the Mother of God with the holy child in her lap is so deeply impressed on Western culture that the inner child is always equated with the Jesus child. This ignores the fact that, just like life itself, the cosmic archetype of the inner child has two faces, one masculine and one feminine. Both are equally important.

In distinction from the masculine principle within us and within creation as a whole, the feminine principle is nonlogical. Thus, the Blessed Mary, as the incarnation of the Western feminine archetype, unites within herself distinct and often even contradictory roles. This was also always true of the much older Goddess figures. If the role of the divine feminine is to hold in harmonious unity all the opposing contrasts of the different planes of being without impairing the differences between them, then of necessity the Goddess must erase the linearity of logic. One can only imagine what trouble the old theologians were put to when they tried to clarify the dissonances in the Blessed Mary's various roles and aspects for the mind to understand. For example, she simultaneously incorporates the divine Mother and the divine Virgin, which is contrary to logic.

There is another way in which the feminine archetype is depicted that the purely intellectually oriented thinking of the modern human finds very difficult to digest. This is where Jesus Christ, the mature son of the Virgin Mary, is depicted as holding on his lap his own mother in the form of a young girl. Then it is Mother Mary who is portrayed as a "child seven days old." This rare image comes from the Middle Ages. It is mostly seen painted in icons representing the death of the Blessed Virgin: Shown below, on the border of

*Detail of a Gothic fresco in the church at Kosec, Slovenia: Christ as the Father is holding on his lap the soul of his mother Mary in the form of a little girl.*

the painting, is Mary's deathbed, surrounded by the weeping apostles. Hovering above is the Christ figure carrying a small girl in his arms. The official explanation tells us that the little girl represents the soul of the Blessed Mary; after Mary's death, the son receives her soul into his arms.

In our quest for the inner child, let us spend a moment contemplating the scene of the Blessed Mary's death. The girl Mary, held on the lap or in the arms of the divine Father (Christ), symbolizes the moment after death when the soul is reunited with the Higher Self, who is represented by the Christ figure. Pictorially speaking, the person's spiritual-soul aspect is thereby raised aloft, so that it will be taken back into eternity after death.

This picture is the counterpoint to the archetypal image of the divine Mother (Mary) holding the little Christ boy on her lap—or in her arms. The "little boy seven days old" on Mary's lap has just been born. He symbolizes our oneness with our eternal aspect, which is still wholly present in the moments after birth.

Translated into everyday language, the two archetypal images confirm that after death, and also directly after birth, one is embedded in eternity and can therefore experience one's wholeness unhindered. The problem lies in the period between—between birth and death—when the experience of eternity is suppressed by the egocentric mind and the influence of social norms. The quest for the inner child offers a path through which we can recover and experience anew the missing wholeness of our own being *without first having to die in order to do so*.

We can now return to our exercise:
- This time, it is Christ as the Father, holding the little girl Mary on his lap, who represents the source of inspiration. Now ask him to hand you the little girl. See how it feels to be holding the archetypal image of your spiritual soul in your lap.
- Then let the child turn upside down and go through the same somersault process as was described above. When she stands

up within you, take particular note of the new and blissful quality that is expanding within your being. Is it distinguishable from how the inner child's masculine aspect revealed itself within you, is it even complementary?

Mind finally gave up on the extreme illogic of its attempt to justify the synthesis of the archetypal roles of the Blessed Mary and Jesus Christ. In consequence, the Tridentine Council, which was held in the sixteenth century, prohibited the portrayal of St. Anne in her true, threefold aspect.[1] Most of the images of St. Anne with Jesus and Mary on her lap were put to the torch at that time; only a few survived the destructive frenzy.

These few are significant because they portray Anne, Mary's mother, holding two little children on her lap: one is the boy Jesus and the other is the girl Mary. How can Jesus and Mary be depicted as children of like age when Mary is at the same time the mother of Jesus?

Here we see the simultaneous presence of the inner boy and the inner girl "seven days old" in the core of the human being. There they are one. The boy symbolizes the person's spiritual identity, the girl their eternal soul.

To better explore and experience the differences between these two archetypal images of our inner self, we can use our imagination to ask St. Anne to hand us each of the two children, one after the other.

In November 2002, I was in Sao Paulo, Brazil, working on the geomantic decoding of the cityscape, when I found that the Museum for Sacred Art there was holding an exhibition of portrayals of the Blessed Virgin gathered from all over the country. It was a great opportunity to show the seminar participants the different archetypal images associated with the principle of the inner child, and together we visited the exhibition. But when we came to the

---

1. The cult of St. Anne the Threefold was widely spread throughout Europe during the Middle Ages. Because she is the grandmother of Jesus the Christ, she was intuitively perceived as the embodiment of the Great Mother—the ancient, universal Goddess. Like the ancient Goddess, St. Anne displays three aspects, for she holds in her hands her daughter Mary and her grandson Jesus, both as little children.

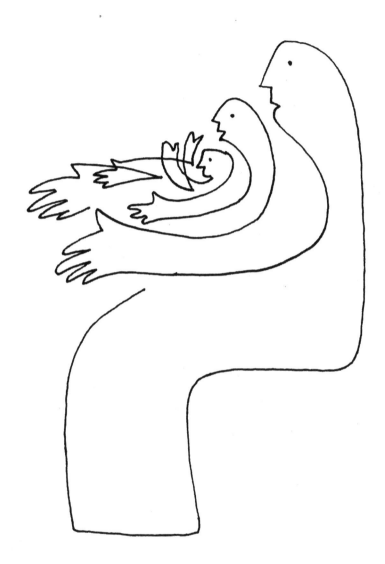

*Based on a sculpture in Sao Paulo: St. Anne the Threefold, representative of the Mother Goddess, is holding her daughter Mary in her lap, who in turn is cradling her son Jesus.*

images of Anne the Threefold, we were disappointed. The pieces on exhibition had all been adjusted to suit the decisions of the Tridentine Council and the particular needs of Mind. Images so derived all show St. Anne as a grandmother: at her side is her half-grown daughter Mary, who is holding the unweaned baby Jesus in her arms—her son and Anne's grandson.

Just as we were leaving the exhibition and wandering haphazardly around the rooms harboring the museum's regular collection, I discovered a small wooden sculpture that showed St. Anne sitting on the "throne of the Great Goddess." On her lap she was holding the Blessed Mary in the form of a young girl. In turn, the little Mary was also cradling a child in her lap, the little Jesus.

Based on the archetype of Anne the Threefold, there is an exercise that can be used to help other people—those who are suffering or have sustained some trauma, or who merely need a loving embrace:

- Imagine you are taking your inner child onto your lap and connecting with it through your feelings. Then, gently stretch out your arms and imagine that your hands are one with the child's tender hands.
- Now you take the next step: Think of the person you want to help and, together with your inner child, take that person's inner child onto both your laps.
- To envision how things are going with the person concerned, you first need to sense that person's presence in your united heart.
- Now the time is ripe for you and your inner child to give the person your joint gift: With your joined hands bless the other's inner child on both the left side and the right. Observe at the same time any changes that can be experienced in the other's inner child.
- After a while, the other person's inner child will wish to say goodbye, and you will hand it back.
- You should then put your hands together in a gesture of thanksgiving.

## The Human Holon

The next two months were spent in intense concentration on the theme of the inner child. Then a truly banal dream drew my attention off in another direction. I had this dream on July 14, 2001. The scene is a construction site. The sun is already high and it has become quite warm. I am wearing a thick jacket, in the pockets of which I normally keep my personal papers, and am about to take it off. First I take the documents from my pocket and put them down on the hood of a humble yellow excavator that is standing nearby. Just then I get a sudden call that I am needed in another corner of the construction site. I run there right away and only after a while do I remember the documents. However, now I don't remember whether I left them on the hood of the excavator or stuck them back in my pocket. I start running back to make certain, but unfortunately the excavator has in the meantime been put back to work and I see it disappearing behind a rise in the terrain. It is clearly impossible to catch up with the machine.

The dream touches afresh on the question of human identity. There can be no doubt that the personal documents are a symbol of identity, but this time the concern is not with the eternal aspect of the human self. The squat, humble form of the excavator that bore the documents away indicates that the concern is now with the change in the physical aspect of the human being. The real question is whether it is only the physical matter of a person's body that is meant, or also one's multidimensionality.

Just as Earth is not merely a material sphere coursing through the universe, so too the human being is not a mere assemblage of bones and musculature. Just as Earth is surrounded by a measurable atmosphere and penetrated by power fields, so too are human beings wrapped in the subtle clouds of their feeling and thinking fields and equipped with a personal energy system. And just as the earthly cosmos is surrounded by protective mantles—the ozone layer, for example—to maintain an autonomous space where it can evolve, so does a protective mantle surround the wholeness of humanity. It represents the membrane of our autonomous space, within which our personal development takes place in the time-span between birth and death. In geomantic practice this autonomous space is called the Holon, which is an expression originating from Greek tradition. The word "holistic," for example, is derived from the same root.

The human Holon, in its multidimensionality, is no less perfect than the Holon of the Earth. One can even say that the Holon of every single human being on the Earth is a fractal—a holographic fragment—of Earth's Holon. In this sense, it is easy to understand why the primitive peoples of humanity describe themselves as children of the Earth. This does not merely express their loving and caring relationship with the Earth Mother, but rather rests on the certain knowledge that the Holon of every single human being is a mirror image of the gigantic Holon of Earth itself.

When I speak of Earth's Holon, I am thinking of her cosmic wholeness, which embraces not only the starlike core of the planet but also the vault of heaven with moon, sun and all the other stars that exert an influence upon her. Between these two poles lie all the visible and invisible planes of the earthly Holon, which make up the unique form of the living planet Earth.

Instead of standing ashamed before Mind's dismissive criticism, we should take pride in daring to believe once again in the old geocentric image of the world. In the geomantic view the Earth is a consummate whole that is centered in its own midpoint. Just like the coarse material layers, so the subtle spiritual planes of earthly existence are organized around this midpoint. In the sense of an Earth cosmos, Earth is autonomous and perfect.

On the nearest plane of power, our planet is a tiny fractal of the much greater Holon of the solar system. This again is a fractal of the galactic Holon. Perhaps the most inclusive possible Holon would describe the concept of God, or Goddess, embodying all that is created and uncreated in a loving embrace.

The idea of a Holon centered in the midpoint of Earth was rejected amid the euphoria of this Mind-imprinted age. It was replaced by the proposition that we, together with all the other living kingdoms of Earth, inhabit a planet that has no midpoint of its own. Mind sees Earth as dependent on a power center far outside of itself—on the sun, in fact. This unhappy estrangement corresponds to the plight of modern humankind, which has likewise lost its midpoint and become a powerless toy in the hands of different ideologies, economic systems and quasi-scientific interpretations of its task in life.

As part of her present process of change, Earth is again at the stage of revealing her own midpoint, and people today should mirror this by turning round and reorganizing themselves afresh around their own midpoint. I use the model of the so-called cosmic cross as the pattern for this inversion of ourselves. It is one of the oldest archetypal images of Earth's various cultures and is depicted with arms of even length enclosed within a circle. The intersection of the vertical and horizontal axes represents the midpoint of the human Holon.

This midpoint may usually be located in the area between the heart and solar plexus chakras, but you may also have the sense of it being situated higher or lower. It is recognized that Eastern cultures tend to center it lower in the belly, while Western ones place it higher, in the heart region.

It can also happen that a person is quite unable to feel this midpoint, or that it seems to have shifted to one side, somewhere alongside the vertical axis that runs down the spine. In such cases it is advisable to work patiently on centering yourself—for example, with the help of the exercises in the Appendix—so that you can at all times find the place within yourself where you feel "at home." This is the point of inner peace which you should never lose, however dramatic the moment in your life.

The human Holon's rounded mantle is represented by the circle that surrounds the center of the cosmic cross. It is often compared to a protective cloak. I am not totally in agreement with this suggestion because, for the Holon, protection is only a secondary function. Its primary role is to create the rounded space within which a life-

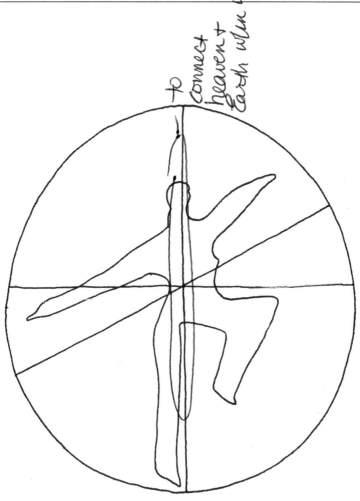

*The Human Holon with its three principal axes.*

*[margin note: primary role of Holon]*

long interaction can take place between the eternal soul on the one hand, and the vital-energetic, emotional, physical and mental extensions of the human being on the other. This is a person's own power space, in which the spiritual self—the inner child—can gradually incarnate, so as to be able to enjoy the fruit of the interplay between the light and shadow sides of life.

This autonomous power space is present in every single human being, and is surrounded by various layers of subtle mantles. They represent the personal protective cloak. On one hand they maintain the integrity of the personal space, and on the other, they permit communication with the surrounding ambient field. One can compare them to the fine membranes that close their tiny pores at need.

The vertical axis of the human Holon runs along the spinal column to connect heaven and Earth within us. Like a heavenly guide, the vertical axis enables our being's spiritual power to descend deep into matter. In reverse, the vertical axis offers the earthly powers of life the chance to soar high into the spiritual spheres of the soul.

Because the human Holon resembles a sphere, there are two horizontal axes in addition to the vertical. Let us look first at the left-to-right axis that connects the Yin with the Yang pole of our being. There is an invisible female pulsating within each physical male, and the male quality pulsates within the physical female also. The Yin-Yang axis makes possible the creative exchange between the opposite poles of our being. Thus, for example, it is advisable to test whether the male-female relationship within your own Holon is in accord before you allow your relationship with your partner to be wrecked by intrahuman problems:

- If you're male, envision that a subtle woman also dwells within you. Her face is turned in the opposite direction to yours. If you're female, it's vice versa.
- Try to give your inner partner as much space as he or she needs: Ask yourself, how does his or her presence feel within me, how am I reacting to it? What can I alter in my life or

way of thinking so that the male and the female halves of my Holon are both happy? They should be able to deal with each other in a way that enriches both.

The task of the second horizontal axis is to connect the two halves of space within me, the light-filled half of space to my front and the dark half of space to my back. Because our physical mind so constantly directs its attention to the anterior half of our Holon's space, we sense the half that lies at our back as being dark, or even absent. Both halves are, however, equally important.

While the anterior space represents the waking consciousness, the posterior space is the seat of the archetypal consciousness and a storehouse of our original powers. When we make our creative plans, we may imagine how their results will look in the anterior space, but it is in the posterior space that we must seek for the causes and powers that are necessary to make them happen.

Today, human beings live almost exclusively in the anterior space. The posterior space is thought of as the box-room of the subconscious and disdained. I, on the contrary, wish to restore the creative exchange between the two halves of our Holon. The most effective exercise that I can offer for this purpose was shown me by the exalted beings of a sacred mountain, the *Hohe Meissner*. This exercise uses the hands, for the hands are a Holon in themselves and are structured like the human Holon. One can do creative things using the front side of one's hands. In contrast, the backs of the hands appear of little use, but, as is true for the whole body, they are no less important. It is the backs of the hands that guide the delicate movements of the fingers.

- Put your hands in front of your chest, placing them so that the palm of one and the back of the other are facing forward and their edges lie against each other. In this position the hands correspond to the Yin-Yang sign, meaning that the anterior and posterior sides complement each other.

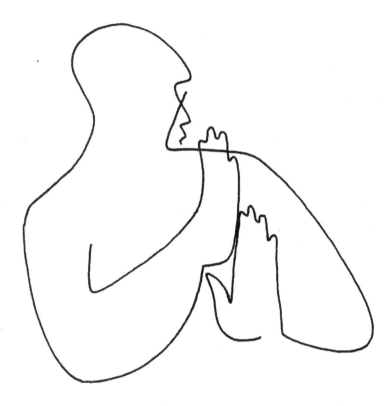

*The Hand Cosmogram for connecting the body's anterior side with the posterior side, which was shown me by the exalted beings of a mountain height.*

- Then begin to circle the edge of one hand round the edge of the other, so that the edges are constantly touching each other. The two hands circle round each other like two millstones. The direction is not important.
- After a while, imagine that the "millstones" are not circling in the space in front of you, but within the space of your breast.

The more often this exercise is repeated, the more it will help to pulverize the blockages that separate the anterior space of present-day reality from access to the storehouse of original powers waiting at the back.

**The Elemental and Spiritual Aspects of Identity**

The impulses for the present process of Earth change are organized in such a way that they are initiated one after the other, but their realization happens in parallel. Thus, I spent more than a year working out all the exercises for the new structure of the human Holon (see Appendix, page 242), but the next impulse for change had already announced itself. It happened during the night after my dream about the Earth bound yellow excavators, and this time the trigger was also a mysterious dream.

In the dream I find myself as a child who has hidden in a room that is half under the Earth. The room resembles a round cellar. I feel quite certain that no one can find me here. Indeed, the cellar has no proper openings but only a sort of air hole to communicate with the outside. Even this opening is overgrown with a profusion of climbing plants and is invisible from without. In fact, the whole cellar space is covered by a carpet of plants, so it is easy to overlook the low-lying building situated in the middle of the forest. Filled with pride at my perfect hiding place, I creep nearer the airhole to look outside. At this moment a boy of my own age stares in through the airhole. Our eyes meet. I am in despair. How can it happen that someone so obviously knows the secret of my hiding place?

No one but our personal elemental being can have access to that secret place within us where the inner child slumbers in hiding. However, the personal elemental being is one of the taboo themes in a culture that, through the mechanism of Mind, constantly demands that its humans distance themselves from the "dark" and "animal" side of their nature. The light of Mind tries to present itself as the unique representative of the human identity. It calls itself Ego. The word Ego carries the meaning, "That which I am (as a human being)."

My dream serves to remind us that there are at least two aspects of the human Holon that may have far more right than Ego to lay claim to the crown of the Self. However, each, in its own way, has been suppressed by the Ego. On the one hand, there is the spiritual aspect of the human identity, which is kept incarcerated in the cellars of the subconscious; and on the other, there is the nature consciousness within us. This last, which is rooted in the heart of nature, has been mercilessly evicted from the personal Holon and banned in the environment, so-called. Nature is certainly allowed to extend its rule around us, but that it should have any part in our identity is, from Mind's standpoint, a heretical idea.

Although Mind may resist the idea most strenuously, from nature's viewpoint, quite truthfully, we humans are a sort of tree species that has learned to draw up its roots and run around in the forest of life's phenomena. And there is more. In a certain respect we resemble the highly evolved animals that have developed the really fantastic capacity of remaining at every moment instinctively embedded in the wholeness of the All. This should not make us ashamed of ourselves, but rather we should be proud of the wonderful gift that we have inherited from nature.

When we ask about the truth of a human being's extensive Self, we should draw into the circle of our close relations not only the beings that are manifest in nature, such as the animals and plants, but also the elemental beings—fairies, gnomes and water nymphs. They incorporate Earth's memory and her ability to consciously guide life's processes. Admittedly, we are not dealing here with a mental, mind-impressed consciousness as with humans. With elemental beings, one is rather speaking of an emotional consciousness that knows no separation. It is a consciousness that thinks of every detail through the embrace of the Whole.

Our present incapacity to think holistically means that it is especially necessary for us to reflect on our relationship with the elemental beings. It would be good for us to understand what we have inherited from this relationship and integrate it into our consciousness once again. To put it another way, one can understand the Ego-consciousness' fear when confronted by our brotherly-sisterly relationship with the beings of nature, because that threatens the presumed superiority of reason in the human world. But, at the same time, the love of the nature beings within us rightfully challenges us to question the Ego's autocratic rule over the framework of human identity.

My dream of June 15, 2001, points to this unrecognized sibling relationship; as soon as the one boy looked the other in the eye, I knew at once that they were brothers. To me, the boy whom Mind had kept hidden in the cellar appeared as if he were flooded with light. The face of his brother, staring into the cellar through the airhole, was shadowed by the leaves of the climbing plants and therefore steeped in darkness. Together, the boys represent our true Self with the light (cosmic aspect) in its relationship to the dark (earthly aspect).

One night as I lay sleepless, the idea came to me of an exercise that would enable us to experience the balance between the two partial aspects of our inner Self. In the dream, I am stretched out on my back in bed when I notice that two boys are lying very quietly within my body—if you are a woman, they would be two girls. The light-skinned boy's scalp reaches as high as my throat. The fontanel of his dark-skinned brother touches my knees, and the soles of his feet reach into my sexual region. There, they rest against the soles of the light-skinned boy's feet. You should make sure you that you get a very precise feeling for the sensitive contact between the soles of those two pairs of feet. This is the paradisiacal touch that makes pos-

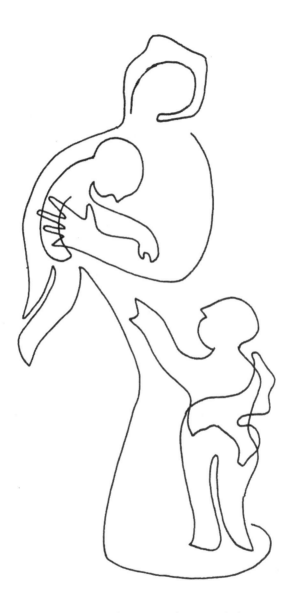

*The two boys, Jesus (The Christ) and John (The Baptist), representing the spiritual and elemental aspects of the human identity—taken from a Renaissance sculpture in the Church of St. Sebastian in Venice, Italy.*

sible the birth of our true Self. Hurrah and alleluia! The new human being is born.

There is a symbol in Western culture that is connected with the archetypal image of the inner child and that points to the peaceful interaction between the two boys. This is the rather rare image where the Madonna is shown with two boys. One sits on his mother's lap, and his hand raised in blessing identifies him as the Christ Child. The second boy stands on the ground and is usually wrapped in an animal skin.

Why should the two boys be depicted like this? Might it be to indicate that with the revelation of the spiritual Self comes the simultaneous appearance of its complementary aspect, raised in Nature?

My thinking points in this direction, but it is not the official explanation. The common interpretation tells us that the second image portrays St. Elisabeth's child, who came into the world at the same time as Jesus and became known as John the Baptist.

A week later I came closer to unraveling the secret of the boy wrapped in animal skins. I had been commissioned by Hagia Chora, the School for Geomancy, to lead a training workshop in Rastenberg in the Austrian *Waldviertel*. The day's assignment was to learn how to get acquainted with the elemental beings of various elements, and we were just on the point of perceiving the beings of a lively mountain stream.

During this exercise a special revelation was granted me after I discovered a beautiful nymph on the opposite bank. Feeling blissfully happy, I watched the play of gentle colors and light-filled clouds composing her body. Suddenly, I was drawn into her "eyes," and the knowledge streamed through me that we knew each other. In the next moment, I realized intuitively that it was not really I that the nymph knew but my personal elemental being. She had even called it by a specific name: "Oberon!"

I sensed that a cloud of feeling, permeated with information, was coming from her to me. From the mass of information imparted, the bits that I could translate on the spur of the moment into my mind's linear understanding had to do with Andersen's fairy tale of the little mermaid: For love of a man, the mermaid allowed herself to be changed into a human woman. The story describes the tortures that the water spirit must endure when she lets herself be shut up in a human body. Lost for her was the colorful freedom of the etheric world.

The nymph of Rastenberg's message was that it was not only the little mermaid that made this sacrifice. Every single elemental being that takes on the task of leading a human child through the gate of birth and is then their lifelong companion undergoes the same tortures. Because people are incapable of orienting themselves within the complicated structure of their physical, etheric and emotional bodies, it may be noted that their personal elemental being remains assigned to their support in every second of their lives. It is a continuing tragedy that our everyday consciousness has completely forgotten this fact. The love for the human, which makes elemental beings so ready for sacrifice at any time, is taken for granted because knowledge of it is excluded from our inner life.

The unexpected eye contact with the nymph of Rastenberg contained the following warning: The time has come for us to surrender our ignorance of our true nature and instead promote the synergy of the two worlds within us. This means, first and foremost, the creation and embodiment of a new image of the human being, one in which both the spiritual and earthly aspects of our identity find a valued place embedded in the multidimensionality of our physical Holon.

**Self-Recognition in the Mirror of Earth Changes**
On July 28, 2001, at the conclusion of the search for the true Self described above, I had a dream which put the efforts surrounding the rediscovery of human wholeness in the context of the changing Earth. You can better understand the dream's message if you know that the house where my family and I live in Slovenia is part of a

group of four buildings that were built quite far from the local town. It takes half an hour to walk to the tarred road where the traffic runs to and fro.

This was my dream: Full of curiosity, I am sitting at home in my kitchen, looking through the window at the preparations for a new railway line that will connect our four outlying houses with the town. The line runs past the other three houses and ends at the edge of the wood behind our house, at the back of which they are just erecting the final stopping point—a very modern construction of glass and steel such as can be seen in big cities everywhere. Already the first train has arrived; it waits for a few minutes to take on any possible passengers. Since there are none, it goes off. I sit at the kitchen window, grieving that this stretch of line has been built at great expense and now the train must go away without a single passenger. It does not even occur to me that I was the only possible passenger and the train was waiting for me.

Translating the information encoded in the dream into everyday language, one can see that it says something about the present Earth changes. The new stretch of railway line is an unmistakable symbol of the absolutely fantastic, and to the mind quite inconceivable, possibilities for humanity's further evolution that are opening up for us through the processes of Earth change.

The concept of Earth change is to be understood as a consequence of the extraordinary alterations in Earth's etheric space that, in my experience, first made themselves noticeable in the autumn of 1997. The term "Earth's etheric space" refers to the planet's vital-energetic sphere, which makes life on the Earth's surface possible in all its dimensions. In the first phase of change, between February and April 1998, the basis of our accustomed etheric space was essentially altered. One can best contemplate the change by using the model of the four elements.

Until now the ruling element has been the earth element. The new "basis" will be influenced by the quality of the air element. One can interpret this to mean that matter (i.e., the earth element) has

### Helpers Stand Ready

Anyone facing such a wide-ranging task as inner change may feel overwhelmed at first by an intimidating sense of helplessness. But you can take comfort! Earth consciousness is offering us an extra-special source of help. This comes from the elemental beings that have evolved through the common consciousness of Earth to help those people who have decided, consciously or unconsciously, to demand and accept that the inescapable processes of change take place in their own inner being.

This gift of the Earth is so unique that I would have passed it by if it had not made itself physically apparent. The scene of the revelation was an isolated village in Portugal that was to be flooded for the Algarve Reservoir, then still in the planning stages. The place is called Luz ("Light"), because in the Middle Ages the Blessed Virgin Mary appeared at a Romanesque pilgrims' church near the village. This old church was also going to disappear in the reservoir.

At the beginning of August 2001 I was invited to Tamera, a co-operative community in Portugal, to take part in an International Peace Camp. Among other things I was to lead a workshop in the area of the future Algarve Reservoir. The purpose of the workshop was to bring the forces of Mercy, Forgiveness and Healing into a landscape that was sorely wounded by preparations for the intended flooding. For example, I saw thousands of ancient cork oaks lying on the ground, felled to make way for what was to be Europe's largest reservoir.

Among other things, we wanted to see whether there was anything that must be done for the pilgrims' church near Luz before that sacred place disappeared into the waters. It happened to be my birthday when we went there, and on the way we had the sense of being accompanied at a distance by a tall figure woven from clouds of light. The summer sky was otherwise bright blue and cloudless. We thought the figure might be a dust devil when we first became aware of it, but the day was calm. Also, the figure's precise threefold division showed that it was no chance apparition but was carrying a message. At its upper end there was something like a spirally rounded head, beneath that a body woven of many strands of light-filled clouds, and below that a train of light which tapered to a point.

Later in the day, when the group was meditating in the portico of the church and exchanging their experiences, one could still recognize the tall figure in the sky. Then, suddenly, it began to go through a process of dissolution. Somebody suggested that if such a light-filled figure had appeared in the same place during the Middle Ages, it would surely have been interpreted as an apparition of the Holy Virgin. This would also fit with the place's name.

I had a further thought. Because the being of Luz had appeared on my birthday, and therefore both on the day when I had been born as *a little child* and at a time in my mature life when I was in the middle of a process of self-initiated rebirth, it must also carry a message for me personally.

I was given the key to the personal message a month later. I was on the little island in the Adriatic Sea where I go each summer to pass my annual retreat. The theme with which I was working at the time was personal change in the mirror of the ongoing transformation of Earth. I was seeking a method by which I could bring my experiences of the human Holon and the resultant exercises into a meaningful whole, so as to offer support to people whom the Earth change had brought into similar neediness and uncertainty.

I am one of those who have chosen to be forerunners, scouting out the path through the labyrinth of change; I gather relevant experiences and share them publicly. I hope that my work will enable everyone to better cope with the tasks that lie before us. Truthfully, this is about a labyrinth of change that in one way or another must be walked in future by all who wish to pass from the old world structure to reach the new Earth.

On that morning I was sitting as usual on the little hill that Julius, my friend and master from the elemental world, has chosen to be his place for thousands of years past. I sent my request for guidance out into the ethers and waited for a reply. What reliable source of support, I had asked, am I to recommend to my fellow humans when their change process begins to falter and they need help urgently? The answer came unambiguous but surprising, that I should turn to the being that had shown itself at Luz.

Now, however, I was several hundred miles distant from Luz and would have felt helpless if the evening before I had not had a meeting that I still could not classify. As dusk fell, I had been going through the lonely countryside to my master's place on the hill when I suddenly felt myself drowning in a wave of fear. I had turned towards where I felt the source of the fear to be and had seen two tall, dark figures sitting close together on the stone field wall over which I must climb whenever I wanted to reach the place where I meditated. My mind told me to draw near and see whether there was not a simple explanation for my fear.

I was stupid enough to follow my mind's advice, rather than orient myself immediately towards meditation and sense whether or not a definite message might be addressing me. When I got near the wall, I could confirm that, in reality, two tall dried-up plants were standing there, which in the half-light had seemed like two beings. My mind dismissed the fact that spiritual beings often use physical events or presences as their medium.

After getting my master's advice, I returned to the two dried-up plants and was able to make contact immediately. Without any doubt, it was the being from Luz that was present—but why, this time, did I perceive two figures? In this moment of uncertainty and doubt, *two* cranes appeared high above me, fell upon each other with loud cries and then, turning in joyful pirouettes, they vanished. My doubts were gone.

Now was the time for mutual recognition. I opened myself to the presence of the unknown beings. A feeling arose as if countless tongues of fire might be dancing and touching me within my body. Yet I felt no heat, only a gentle curiosity.

Next it was my turn. I took a few steps back, to look at the phenomenon "from the outside." It was perceptible as a ball of light

*The twin beings, which are angelic and elemental beings intermingled, stand ready beside humans to help us through our changes.*

strewn with triangles whose points were raised upwards and out. Further examination showed that this was no landscape angel; their presence in the landscape is held anchored by vertical pillars of light. In the case before me, the resemblance was much more like that of an angel in the sense of the Greek word *angeloi*, which means "messenger." It was not only the dynamic form of their ball that led me to this conclusion, but also the merry way they had with each other, full of laughter and jests.

I used the close contact that had meanwhile arisen to learn more about the tasks that the "twins" had undertaken: They were a sort of mingled being, put together from an angelic and an elemental aspect. They were two and yet one. This union had become possible after certain strands in the evolution of elemental beings had undergone a fundamental change.

This has been occasioned by a transfiguration of the elemental world through the Christ presence, the story of which is told in my book *Daughter of Gaia*. This has brought many groups of elemental beings to the expression of such a high degree of love and sympathy that they have even become capable of incorporating the angelic quality. This means that the quality of the angelic world as well as the forces of the elemental are to be felt within a twin being such as I had got to know through the revelation at Luz. In these beings the cosmic and the earthly are woven together.

The twin beings are attracted individually to those people who have the courage to throw themselves into the surges of inner change. However, they may not involve themselves unbidden in one's personal processes. They are much more inclined to accompany the relevant individual from a distance and wait patiently for a time when conscious collaboration becomes possible.

I also asked the twins how someone could begin a conversation with them. For this they gave me two signs. First, they touched my earlobes. Then, after a little while, my tongue was raised and pressed back as far as possible.

There are two tiny chakras of the fire element in the earlobes, and the twin being, as described above, is indeed close to the fire element. Touching both earlobes therefore serves as a recognition code for the twin that is being called. Bending back the tongue is in the same category, for symbolically it represents change. It stimulates a certain resonance with the process of inversion, which is a preparation for the new structure of Earth and humanity.

It seems that the twin beings' task is to provide help with the new configuration of the human Holon. The procedures described below can be used to make contact with your own helper and ask for its assistance with whatever change you are directly facing:

- Go into your inner silence and make sure that your mind has taken up a stand-by position.
- Now touch your earlobes. It is best to rub them lightly for a while between thumb and index finger.
- Now bend your tongue backwards for a moment and open yourself up to the space that has its source behind your back and from there extends outward to all sides—and to the front too. Everything is now ready for a conversation.

CHAPTER 2

# The Fundamental Causes Leading to the Catastrophe of September 11

### The Unexpected Gift of Catharsis

MY ENCOUNTER WITH THE TWIN beings of Luz took place five days before the world-shaking catastrophe of September 11. You will know how a group of highly trained Moslem terrorists hijacked four passenger aircraft almost simultaneously. Two of these were flown, one after the other, into the two skyscrapers of the proud World Trade Center in New York: The result was inconceivable destruction.

After the crime my first reaction was to sit down and quickly write a book to help mediate any hostility between Muslims and Christians, for such was expected to ensue. However, no book of mine ever came out of that horrific deed. Today I am certain that the twin beings had reappeared on that evening in order to draw my attention away from the superficial connections leading to the catastrophe, and so open my subtle senses as to the real causes.

I was first aware of the meaning behind this fortunate shift of focus when, exactly three weeks after the catastrophe, I stood in front of the ruins of the twin towers of the World Trade Center. I gazed astonished at the gigantic heap of savagely interwoven iron bars, crooked balks of steel and concrete rubble. A flock of yellow excavators was scurrying hectically around the pile, attempting to erase the shameful monument as quickly as possible. Suddenly I remembered when, 11 days before the event, a dream had shown me a similarly frightful image.

In the dream, the story begins with the banal need to find a parking place. I am in front of a tall office building where I have some urgent business, but can find nowhere to park my car. In my haste, I decide to leave it for a few moments on the steep ramp leading up to the tall building, although I know that it must be forbidden to park there. I make the hand brake firm, but forget to raise the window. And then I am off.

While I am away a man whom I know in everyday life comes up to my car. He bears the Slovenian version of the common Muslim name "Suleiman." It does not escape him that the car window is open. Unnoticed, he reaches in through the window for the hand brake, releases it, and makes off. The car begins to roll slowly backwards and finally leaps over the edge of the road into the depths below.

When I come hurrying back, the car has vanished. I peer over the edge of the gulf and see the scene changed to a great pile of wreckage: just such a pile of skeletal shapes of iron and steel and wrecked structures as I saw in New York on October 3, 2001, when I stood on "Ground Zero" in front of the ruins of the World Trade Center.

The dream does not need much detailed explanation. It told me what was about to happen: a catastrophic attack in which evildoers of Muslim background would make smart use of inadequate airport security measures. Even the assault weapons—just think aircraft!—they would cold-bloodedly turn against themselves. This lightning-swift action brings with it inconceivable destruction. I must add that the information contained in the dream was so cleverly veiled that it was not possible for me to make any sort of public announcement. As long as the catastrophe had not taken place, I could not know what country and what building were concerned.

The real message in this dream of August 31, 2001 was to deny that the approaching event of 9/11 could be some random occurrence. Here I must mention that much earlier I had had a message about a coming event filled with unfathomable tragedy. In the night of August 11–12, 2001, immediately after the joyful revelation of the twin beings of Luz when I was visiting Portugal, I had a most disturbing dream. Its content did not fit at all with the personal transformation process described in the preceding chapter.

In the dream I am living with my family on a plateau overlooking a mighty city. Suddenly and without any sort of logical reason, my grown-up daughter begins to weep bitterly. A few moments later her tears become inconsolable lamentation. Now she begins to move down a path towards the great city. The nearer she gets to it, the louder and more shattering become her howls. I run after her and try to comfort her and discover why she is lamenting so. Despite my efforts the cause remains unknown.

From the perspective of the catastrophe on 9/11 the dream's message can easily be understood. It simply informed me that something was going to happen to humanity for which our ordinary waking consciousness was completely unprepared, and the impact of its tragedy would burst apart the framework of all our hopes and expectations.

It quite often happens that an unexpected event will spring out at us, without any prior warning. This does not mean, however, that the event was not being prepared long before, vibrating on the unseen planes of Being. The murderous attacks of 9/11 were planned in secret years before they suddenly appeared on the visible plane, sowing their seeds of death.

This means that in the night of August 11–12, when I had the first dream, the catastrophe of 9/11 was already fully underway. In an ideological and logistical sense it was already nearly fully formed, like a child in its mother's body shortly before the birth. It only lacked a final physical shape.

When one has grasped the multidimensionality of life, it is not hard to understand how the embryo of a future phenomenon can easily be changed, provided that it is still passing through the "before-birth" phase and has not yet solidified into physical form.

This has to do with the "phase of the etheric formation" of a happening. This is the phase in which the future event takes on its shape. Its form is still soft (etheric), such that it can be molded differently.

Who knows how often the many missiles of misfortune, aimed at us from the unseen realm, are deflected from their goal in this final etheric phase, before their fateful physical appearance? Who causes that? I believe firmly in the mercy of the spiritual world.

Judging by the dreams I have described, the phase of the etheric formation of the 9/11 catastrophe lasted exactly 30 days. This also defines the time period in which it would have been possible, spiritually, to turn the catastrophe around. Why did this not happen?

In all honesty, I must admit that I did initiate such an attempt, but unfortunately it accomplished nothing. In part, one can attribute its lack of success to my own consciousness, which was insufficiently clear about the preceding steps in the phase of the event's etheric formation. Only now, at a distance of more than a year, can I put the individual messages about the approaching catastrophe in their correct order.

The dream of August 11, featuring my lamenting daughter, indicated that the coming event was starting on its phase of etheric formation. The second dream, where the car fell off into the abyss, marked the high point of that phase, when the catastrophe had begun to take on physical form. From that point in time there were still 11 days in which to work against the process leading to its physical appearance.

In this connection I must make clear that by "working against" I have in mind no semimagical operation, in the sense of a manipulation of the life processes. No, I am thinking instead of the human capacity to work cocreatively in shaping reality. I am deeply convinced that this capacity is also our calling and is part of the intrinsic pattern of the gradually unfolding Being of Humanity.

When I dreamt of the crashed car amid the ruins, I was visiting Eberharting in Bavaria and engaged in a workshop under the auspices of the Hagia Chora School for Geomancy. I did not get back home to Slovenia till September 3. As an initial step, I called up a dear friend in Switzerland, asking him to help me put out a worldwide call for a group meditation as quickly as possible. For several years now the "Life Network of Geomancy and Transformation" has been called into being for just such purposes.

Without delay I sat down at the word processor to write out the message for the meditation. The machine refused to work. I felt as if disabled. Something similar happened to my friend in Switzerland. In consequence we decided not to proceed with our plan.

If matters had unfolded differently, I might perhaps have proposed the following exercise that harks back to the archetypal image of Christ holding a little girl in his lap. She is the Blessed Mary, ever Virgin. Christ represents the Core of the Universe, and the Virgin incarnates the World Soul through whom the Core of the Universe works to protect and nourish creation and impel it forward. The exercise is as follows:

- Center yourself in your midpoint; you are rounded off within the mantle of your Holon.
- Think of the process on whose transformation you would like to work, and envision it as a ball of light hovering within you. The size of the ball will adjust itself of its own accord. Make sure that all the aspects belonging to the relevant process—energetic, feeling- and form-related—are present within the ball.
- Now gently open your hands and ask the Core of the Universe to reach across and give you the little maiden. Let her Being slowly glide through your body until the World Soul is fully present in the realm of your back.
- Next direct your attention to the ball of light once again. Imagine that the presence of the World Soul behind your back is gradually dissolving into a spray of countless drops of water that continually stream through the light ball to release the processes of transformation within it. With the help of this penetrating stream the contents of the light ball will be

brought into accord with the harmonies of the Cosmic Whole.
- Now put your hands together and give thanks. This is also the signal for the presence of the World Soul to take its leave.

Let us now return to the question of why it was not possible to work on transforming the event of 9/11 before it took on its horrific form. The answer to this question was shared with me through a parallel layer of dreams. The purpose of their message was to reveal deeper connections behind the assault on the twin towers in New York.

The first dream takes place in a dining hall so immense that its walls are out of sight. People from every nation are sitting in a long row on the floor, waiting for their meal. A personality of unbelievable strength and radiance is sliding on his knees from one person to the next and asking each one exactly what he wants for his midday meal. However, I have the sense that the process involves a very weighty problem that remains unresolved.

I had not succeeded in identifying the problem before it became my turn. Now the exalted personality is kneeling in front of me, holding a broad dish in his hands. On the dish are samples of all the wonderful foods that the master has to offer. Most of them are unknown to me.

For my part, I too am holding a plate in my hands, full of various leftovers from the previous day's meal. When I look at the new foods offered me, I become terribly nervous and undecided. Within myself, I feel drawn to the new foods. At the same time, I am not ready to forget the old leftovers. Instead of accepting the offer of the new and letting the old go, I begin to move the leftovers around on my plate with a fork in the hope of making some space for the new foods. Inside myself however, I know that my attempts will have no useful outcome.

Still doubtful, I decide to heap the leftovers into two separate piles. And there the dream ends. Looking back from the standpoint of the catastrophe of 9/11, one can see the significance of that final image with its two piles of food. It is a sign that the dream refers to the tragic story of the twin towers of the World Trade Center, although I received it on August 1, 2001, well before the event.

When I awoke, I felt confounded. My mind reproached me bitterly for holding fast onto old projects and leaving no place for the new concepts that wanted to manifest themselves through me. My intuition, however, had a different opinion. It felt that the dream was not about my personal problems but the problems of our world civilization. The dream was composed so that I could sense the threatening difficulties of the world situation within my own body.

The causes that led to the catastrophe of 9/11 could be recognized very precisely in the atmosphere of the dream. I sensed the dramatic conflict in which human civilization is caught at the present time. On the one hand, the Earth changes described above are offering humanity fantastic possibilities for further evolution; on the other, people are clinging fast to the old, well-traveled patterns, with their bipolar structures and hierarchical models. And so we lose the free, open space that is unconditionally necessary for the new possibilities to unfold.

Our basic indecisiveness is holding us in a state of unbearable tension, and this acted like an irresistible magnet drawing the catastrophe of 9/11 to us. Properly speaking, we should no longer talk of a catastrophe but rather of a message, even an educational experience.

That message found still clearer expression in a second dream. It was imparted on August 30 at Eberharting in Bavaria, 12 days before the two aircraft crashed into the twin towers of the World Trade Center. The story of the dream tells of a wedding already in progress.

The dream is somewhat complicated, and I ask for patience. It started as follows: Arrangements have been made for a bus to take the wedding guests to the place where the wedding breakfast will be served. The father of the bride and half the guests are already gathered at the bus stop, but the other half of the guests, together with the bride's mother, are still missing. We wait and wait.

Finally, the missing guests appear, but the mother is not with them. A woman who is part of the group that has just arrived comes up to me and explains the problem. An electric heater had been ordered to warm the room for the wedding guests, but the bride's mother received a phone call to tell her that it cannot be delivered in time. The news has affected her so badly that she has suffered a nervous breakdown.

The woman's explanation seems straightforward, but contained a play on words that is critical for understanding the fundamental causes of the catastrophe of 9/11. When speaking to me, she used an expression that coupled the word "elec(tric)" with first name of the supplier of the heater, "Tron(telj)." This produces the word "Electron," and this points to a problem in the vital-energetic field.

At this point the meaning of the dream becomes very difficult to unravel, since it obviously concerns a very complex statement. Somehow aware of this, though still in the dream, I decide to leave the wedding guests and seek out the bride's mother, to learn the truth first-hand. Unhappily, I do not know the way. I wander around in the town, but there is nobody in the streets to give me directions.

Finally a little old lady emerges on the opposite side of the street. Joyfully I run towards her to ask advice, but at the next moment stop bewildered. I do not see a woman now, but a pile of newspapers that are moving themselves on two legs. At this moment the pile begins to change. Layer after layer of newspaper is shoved aside, and soon I am looking into the eyes of a woman who is small, broad and immensely old and wise.

Never before have I experienced the Earth Mother so close to me, near enough to touch. She is ugly, her figure squat and broad as a truck. At the same time she is beautiful, and a loving and humorous grin gleams in her face. With a simple wave of her hand she points in a particular direction. Immediately afterwards I find the bride's mother.

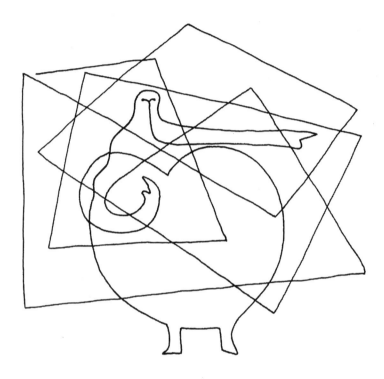

*In my dream of September 11, 2001, Mother Earth appeared wrapped up in countless sheets of newspaper, and showed me the way.*

My worst horror was still to come: the mother was hollow. I was at my wits' end. I did not know what else to do but hold her fast against my heart. It was useless. Put simply, she was void. She had become the symbol of "the second death."

The Gaia figure, wrapped in those endless layers of newspaper, stands for Earth, which is suffering under humanity's repressive, reason-directed control. The common consciousness of Earth has become wrapped in inconceivably multitudinous layers of mind-conformable systems through the globalization of our rationalistically oriented civilization. I am not referring just to the countless electronic networks engaged in data exchange, but above all to the desiccated, mind-conforming ways that prescribe how our life processes are devised and actualized. Our usages of atomic power, and now gene manipulation, have finally stamped a mental corset on even the subtlest levels of the manifested world.

On the other hand, the transformation of the planet's subtle body, taking place within the framework of the current Earth changes, has advanced so far that the new ordering of the Earth cosmos is ripe for external manifestation. This offers a real possibility to heal the Earth and return humanity to the rhythms of the cosmic harmony. In picturing the wedding preparations my dream suggested this happy outcome.

Despite this opportunity, humanity remains stuck in the narrow mental structures that rob people of every trace of sensitivity for the epoch-making changes taking place around them in Earth and nature. This has caused so great a polarity between the power fields of Earth and humanity—called "electron fields" in the dream—that it threatens the whole field of Life with total breakdown. As regards us human beings, that would deliver us into the danger of "the second death."

The concept of "the second death" appears in the Revelation of St. John and denotes the complete dissolution of one's spiritual and soul identity. "The first death" represents the natural transition from one dimension of being to another. People certainly lose their physical body, but continue the further course of their lives on the spiritual plane. "The second death" describes the final extinction of a person's memory, so that the cosmos can no longer recall that he or she has ever existed.

We shall not come to such a pass. To protect us and save our desperate situation from the threatening abyss, the devilish plan devised by the Al Qaeda terrorists was modified in the phase of its etheric formation. Divine mercy, acting in the period between August 11 and September 11, planted a very positive and powerful seed in the approaching act of destruction.

Does such a concept not justify the terrorists' atrocity and set a most dangerous precedent for the future?

In no way! Criminality remains repulsive criminality. The perpetrators must bear full responsibility. Throughout the world, the codes of law and associated courts of justice have made this clear. The idea that good may come out of evil does not justify the evil. Rather, the idea of a saving grace grows out of the experiences described, and it is my conviction that indwelling in the divine mercy is such pure power and wisdom that it can shine even through a crime and transform it to an act of redemption.

This statement is supported by a passage in the Revelation of St. John, Chapter 17, verse 17, which runs as follows:

For God the Father has planted in their hearts a purpose such that in the end they must act according to his mind. Thus, they are also serving in a (divine) sense even while they place their kingdom in the service of the beast, until the goals of the Words of God shall be fulfilled. (Translated from *Bibelausgaben, Das Neue Testament, Originalfassung* by Emil Bock.)

I should make clear that "they" in the above quotation refers to people who cherish false ideas about the meaning of life, or who follow destructive goals. Among them, by my understanding, are the planners and operatives who carried out the crime of 9/11. They are those who "serve the beast."

When I visited New York at the beginning of October 2001, I

found confirmation of this immensely encouraging ability to reverse the poles of good and evil. As mentioned, on October 3 I visited "Ground Zero," where the towers of the World Trade Center had stood. Not allowing myself to be absorbed in the prevailing tide of common curiosity, I instead paid attention to the place's soul quality. To my surprise, I had a sense of overwhelming mercy extending outwards from the place of the tragedy. Seeking to establish how this beneficent feeling could be there, despite the devastation of the ruins, I opened the eyes of my soul.

What I perceived was a light presence—to my soul sense it formed a pillar as tall as the two towers had been—and from it there flowed the quality of mercy. Within this pillar of mercy I spontaneously recognized the presence of the Virgin Mary, who in my consciousness can also be Kuan-Yin in the Chinese tradition, or the Buddhist Tara. Her quality of compassion radiated authority, surrounding and streaming through the whole area.

It was two months later, when I was in Cologne conducting a seminar on the theme "Self Healing for Earth and Humanity," that I came nearer to understanding the meaning of the pillar of mercy. We were working on an exercise which I call "The Tear of Mercy" and which originates from my last book, *Daughter of Gaia*. In this exercise we select a place where there is need and suffering. Next we raise our hands to heart level to form a horizontal channel. Then we pray to "the Divine Virgin" (who is the Soul of the Universe) for a tear of mercy, which our imagination guides to the place selected.

One of the seminar participants suggested that the group guide the tear of mercy to the ruins of the World Trade Center in New York, because she felt that many of the souls of those who had died would still be suffering there. Her proposal was accepted and for the next few moments the group's attention was concentrated on the site of the attack in New York.

During this short ritual I had a vision so overwhelming that I could not bring myself to share it with the rest of the group when later we exchanged our experiences of the exercise. I needed more time to fit the vision into the overall picture.

This is what I saw. At the moment that the tear of mercy touched the site of the tragedy, the light presence—i.e., the pillar of mercy previously described—lifted something like a veil. For a split second I was allowed a glance into the abyss of death that the mass murder had opened and whose poison the veil of mercy was mitigating. The dimness of death that I perceived in that place was deeper than the darkness of the blackest night. I associate this experience with catharsis.

In a Greek tragedy, the hero, usually a much-revered figure, is not actually slain. A virtual killing is enacted, and yet the audience in the amphitheater experiences the scene as a real encounter with the abyss of death. This encounter with the tragedy of death affects the audience so deeply that their feelings are deeply shattered and they undergo an inner change. Such a transformational experience is called "catharsis." People who have experienced catharsis emerge from the theater as if newly born.

To better understand the foregoing, the reader should know that the "Abyss of Death" is identical with the pulse beat of eternity. Human beings must leave the dimension of eternity to become born into the world of matter, but in the natural order of things they regain it in the moment of death. From the viewpoint of the incarnate world, the moment of death plunges a person into an abyss of darkness. From the viewpoint of the soul, the same process is reversed, and—as described in near-death accounts—it is experienced as a blessed dissolution into heavenly light.

Extending this thought to the tragic events of 9/11, it means that millions of people worldwide then experienced the closeness of death to such effect that the thick layers of mental consciousness into which, like a corset, we have been forced, were for a split second torn aside. In that moment we all saw the face of eternity. Although this did not happen consciously—for our waking consciousness was plunged in chaos and extreme uncertainty—it has anchored the

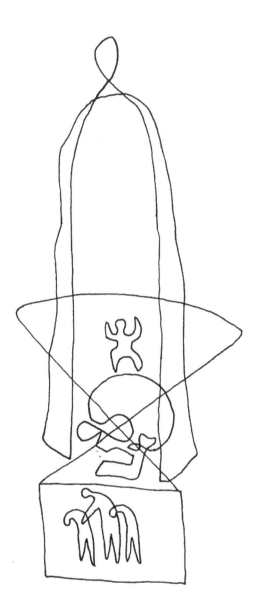

*The figure of the Goddess of Mercy, who has wrapped her cloak around the ruins of the World Trade Center, has revealed to me the secret of catharsis.*

experience of perfection in our very bodies so that it can nevermore be erased.

Its effect was very similar to a Greek tragedy. We did not all have to die in order to experience the liberating catharsis. A relatively small number of our fellow human beings sacrificed themselves on behalf of countless others to achieve the reconnection that the times demand. And we, the fortunate recipients, were not sitting in a Greek amphitheater but in front of our TV and radio sets.

This accomplished two things. First, our momentary contact with eternity enabled humanity to reconnect with the original ground of our Being. The danger of "the second death" described above was thereby successfully overcome. This also opened the way for the continuing transformation of Earth and humanity. I feel deep gratitude to those who sacrificed their lives so that we could live on.

Second, another wonderful thing happened. Enormous quantities of the negativity that human beings had accumulated within and around themselves were burned away in one single moment by the terrifying onslaught of divine power and perfection. This negativity was mostly about our denial of the true Self, a burden that humanity has let grow ever heavier through our ways of thinking, feeling and commercial dealing. We have no conception of how completely everything about us was thereby transformed and redeemed. For this too I feel deep gratitude. The dead deserve all honor.

### The Powers That Work Against the Earth Changes

The catastrophe of 9/11 resulted in yet another decisive turn in the course of the Earth changes, for which too I am indescribably grateful. The flood wave of soul-affecting shock released by the crime burst so powerfully on the emotional plane of the world's fabric that our eyes were torn open, and for a moment the hidden causes of our chaotic world situation lay naked before us.

This meant that the forces that had until then worked in secret against the transformation of Earth and humanity were forced to show their true colors. This was confirmed in a dream I had on

September 22, 11 days after the event. In the dream, I find myself in dirty, oily water, observing a duel between two "giants" who—standing up to their knees in water—are engaged in a life and death struggle. A radio reporter is standing on one side and excitedly shouting a commentary on the battle as if at a football match!

The fact that I am observing the duel while deeply immersed in water should be sufficient confirmation that, alongside the visible plane of world events about which the international media comments unceasingly, there exists another plane. This plane is concerned with processes within the emotional-energetic or astral field of humanity, which is ruled by the water element. We do not usually perceive this plane and we therefore pay it too little attention.

On this plane dwell the powers that have arisen by reason of our egocentric drives, and from within it they craft the complex causes of world events—causes that we never think to suspect. What glimpses we see are mere isolated happenings that suggest no deeper interconnections. The events that show up from time to time on our TV screens, isolated from their causes, are really the outcome of decisions that are never recognized.

After a while the two struggling "giants" come to the moment of decision, which in my view may be equated with 9/11. In great excitement the commentator informs us that one of the combatants has been mortally wounded, and calls out his name. It is "Bosko Hodzic," a common name among Bosnian Muslims. After that there is absolute silence.

I now emerge and see a modern man sitting at the edge of the water where the struggle took place. I know him to be modern because he is wearing a suit that has come straight from the shop. He feels in his pocket, brings out his glasses and puts them on his nose. I sense how amazed he is that he can suddenly see. Of the two combatants, there is now no trace on this plane of external existence.

To better explain the dream's message to our waking consciousness, I suggest we should explore the roles of the two "giants" who were tangled together in merciless combat. They stand for the two forces that, in the 9/11 catastrophe, exploded against each other. Together, they represent the power that opposes the transformation of Earth and humanity.

These contrary forces are not basically a power that is evil in itself, and as such tries from the outside to influence world evolution with destructive intent. The catastrophe of 9/11 has revealed these contrary forces to be two aspects of modern humanity and its civilizations that have become separated from one another.

One of these forces is expressed by the immeasurable power of capital and the global economy, represented by the World Trade Center in New York that is now destroyed. In respect of this force I am thinking of the merciless exploitation carried out worldwide by global economic enterprises that have as their goal the enlargement of worldly power and material wealth, which inevitably leads to individual and cultural despoliation. Here, the driving force is a type of consciousness that calls exclusively on a person's intellectual capacity and thereby feigns mental purity while emphasizing the logical comprehensibility of its own projected model. Thus it hides the fact, so obvious to the feeling heart, that its encapsulated and egocentric consciousness has become estranged from the wholeness of life and is causing extreme emotional suffering to those who are both its bearers and its prisoners. The pain is hidden by the scientific and technological advances that fascinate us and at the same time lead us into still deeper estrangement from ourselves. Cybernetics is but one example.

On the other hand we have to deal with people who have surrendered themselves, their devotion and their deeds, absolutely and exclusively to God, and in this mindset see themselves as warriors of God—for example, the terrorists of Al Qaeda or Al Fatah. They do not notice that their devotion relates to a God that has arisen from their own projections of the divine, and that their God's particular qualities are defined by their own intentions. Here we see, as in the polarization of opposites described above, a trick of the intellectually oriented thinking that is the brand mark of modern men and women—humanity in the Age of Iron. Modern human beings cannot decide whether their religious affiliation relates to the Source of All Being, or to the projected model of the Source that "their" religion has established within the framework of its theology.

To be fair, I must emphasize that the same conflicted mental estrangement underlies not only Islam but Christianity and Judaism too. All three world religions are children of the "Age of Iron."

We should also realize that reason's control over us is no modern phenomenon but was encouraged by the Enlightenment of the eighteenth century. In fact, we are dealing with a development from which only the so-called primitive peoples can withdraw, and they only insofar as they remain untouched by "civilization." This development has been rolling forward for the past 3,000 years of the "Age of Iron," and its goal is that we should gradually learn how to make our inward independence real in our external lives.

There is a gigantic shadow associated with this "Age of Iron," one that we have accumulated by misusing the qualities inherent in our freedom and independence. In the final analysis it is this shadow, generated by our human evolutionary process, which gives rise to the contrary force working against today's essential transformation of humanity and Earth. In the thirteenth chapter of the Revelation of St. John, this contrary force is presented as having two separate aspects, exactly as revealed through the explosive collaboration that resulted in 9/11. Though separate, they incorporate one and the same principle and cooperate with each other. On the one hand St. John writes of a beast rising up out of the sea, "That had ten horns and seven heads, and on the horns were ten crowns and on its heads were names that blasphemed against the Spirit of God." On the other hand he writes of the aspect symbolizing the false prophets that spring from the solid earth: " . . . and it had two horns so that its aspect was that of a lamb, but its speech was that of a dragon."

For all of us on 9/11, it was a powerful shock for the two polarities of our normal schizoid and fragmented consciousness to crash into each other and so reveal themselves—independent of whether we afterwards consciously absorbed the stunning message or simply allowed it to affect us on the emotional level. The crime's astonishing mixture of cunning, force and considerable technical knowledge testifies that the two polarities were cooperating with each other, either consciously or unconsciously, to mutually inflict a mortal wound.

Let us return to the dream about the battle between the two giants. Its final sequence contains a message that the world experience will from now on incorporate a new dimension, which we will overlook at our peril. I am thinking of the dream's closing image, which portrays the modern human being who is trapped in his intellectual consciousness and sits isolated on the edge of the external world. He has no inkling of what is happening in front of him under the oily water—in the invisible astral realm. There, a battle for control of the world is taking place. The completely divided polarities of our modern consciousness are fighting doggedly against each other to obtain control of a world *that is not there anymore.*

How can the world no longer be there when we see it all around us?

We have suspected it for a long time, but we all shy away from the truth of it: The ever more intense concentration on linear time, driven to extremes in the last few centuries, has caused us—and our extensive, worldwide civilization—to lose the last vestiges of contact with the unfolding reality.

As outlined above, Earth is transforming herself and has already abandoned the basic structural plan to which we are accustomed. She has built herself anew. Now she is poised to lead the quality of time and the expansions of space into a new dimension. (Post)modern humans, however, remain encapsulated in their own projections of space/time, which evolution has already dismissed from its purview. Is this situation not extremely perilous for the very being of humanity? Not for nothing were we warned of the danger of "the second death."

In this context it is plain that soul-shattering events like 9/11 can be a help and a blessing. They manifest themselves in a distinct rhythm in order to shake humanity up and drag it out of its encapsulation in a false time frame. That representative of modern human beings who put on his glasses at the end of my dream was really telling me that these unhappy events help him inwardly to reconnect with reality. And so, renewed, we should express our gratitude to the dead and give them honor.

On September 20, 2001, nine days after the crime, when I had to choose whether or not, in light of the dangerous world situation, to fly to the USA, I was given a dream that took away my fears. Its message relates to the crime's disturbing aftermath, counting in the sharpened controls on the processes of ordinary life, the curtailment of civil rights and finally the war in Afghanistan.

In my dream, a great eagle settles itself on my head. To my astonishment however, it is as tame as a chicken. It makes itself small and presses as closely as possible to my head. In doing so, it lets its wings slip down so far that they threaten to obscure my sight. I must constantly push its feathers aside in order to see. However, the eagle shows no signs of ever wanting to leave my head. My best course is to lift it off. Its talons are thrust in so deep, so deep! It is a fact that my scalp hurt for a long time after I had awakened.

Translated into logical language, what is the dream's message?

When their efforts to prop up the old world model resulted in the mutual explosion of their polarities on 9/11, the forces opposing the transformation of Earth and humanity lost much of their real power. It is certainly possible that they will still make more consistent demonstrations of their pretended superiority. However, it would be a mistake to pay more attention than necessary to such a show of force. It makes no sense at all to be overly frightened of those forces that continue to promote the old world model, or to get involved in fighting them. Or, to put it a different way—these opposing forces can only maintain their presence here because the majority of humans believe in their existence and either revere or fight against the symbols of their power.

What alternative is there? As you work to resuscitate something,

## The Fundamental Causes Leading to the Catastrophe of 9/11 / 65

*An eagle, America's heraldic bird, its back broken, alights on my head and digs in its talons hard.*

how can you gradually withdraw its compulsion to be something that no longer exists?

There are various possibilities. It is left to the intuition or creativity of each individual to determine his or her course in any given moment. There are no hard and fast rules. Here are my recommendations:

- We should learn to listen to the voice of our own eternal soul and to follow its guidance. It reveals itself through the holographic language of events that unexpectedly transfigure or perhaps shatter our lives. It reveals its message in dreams in the night or intuitions in the bright daylight. Its voice is perceptible in the heart's impulse or in the inner silence. To prevent another catastrophe like that of 9/11, we should learn to give ear to the voice of the soul—which is the voice of eternity within us—and not suppress its blessed presence in our personal and public lives. For this purpose, the preceding chapter proposed various exercises based on the archetypal image of the inner child.

- We should, as always, work on our own reconnection and integration with our personal Holon. From page 237 onwards, the Appendix proposes a sequence of exercises that can help nurture one's own wholeness in harmony with the archetypal pattern of the cosmic Holon of humanity. It is most important to find one's own midpoint and remain anchored there, irrespective of what is happening around one. The catastrophe of 9/11 teaches that future circumstances may arise that hurl the masses into emotional and energetic chaos. The catastrophic consequences for Earth's equilibrium and our personal stability can only be avoided if at such moments there are a large number of individuals worldwide who maintain their inner silence and remain centered in their midpoint.

- Apart from this we should exert ourselves on the progressive synthesis of emotional and mental powers and intentions. We may picture this as renewed communication between head and belly that finds its synergy in the heart. It represents the synthesis of heaven and Earth within us. If we gradually become capable of loving through our thinking and imagining through our feelings, we shall bridge the deadly gulf that drives us to suicidal acts like that of 9/11. One is then grounded in the force field of one's own heart and the sense of what is true and untrue. One is empowered to preserve and protect Life.

CHAPTER THREE

# After the Catastrophe: Humanity's New Inner Organization

**The Cosmic Double**

I AM CONVINCED THAT THE past five years have been the most momentous for the future evolution of Earth and humanity, in which connection I am referring also to the change in genetic codes. This means that during the years 1998 to 2002 the influence of an unknown force so fundamentally altered the archetypal patterns of Earth and humanity that our further evolution could take a quite different road than the one we project for ourselves today.

However, my intuition is dissatisfied with the term "change in the genetic codes." To convey a wider meaning, it proposes a quite different type of language and conceptualization to accurately describe the message of the last five years. We may compare these years with the public and historically attested actions of Jesus Christ, which only lasted three years but completely altered the course of world history. Yet the person and astounding deeds and teaching of Jesus Christ remained practically unnoticed by contemporary historians.

In my book *Christ Power and the Earth Goddess* I tried as far as possible to separate Jesus Christ's influence on the evolutionary history of Earth from its too narrow association with his historical personality. We learned from this that Christ represents a force in the cosmic consciousness that is devoid of any personality features and can therefore encourage the evolution of humanity and guide us in the direction of inner freedom and individuation. We should not think of this force as connected only with the name of Christ. Other cultures may use other names for the same cosmic force. Buddhists, for example, reverence it in the figure of the Maitreya Buddha. What we are contemplating here is a subjective-objective force of the universe that—like a star—approaches Earth and humanity again and again to influence the key moments of our evolution. Its aim is to change and renew the inherited archetype of our evolutionary pattern, which is our genetic code. It thereby releases new impulses in momentous ways.

Humanity has developed different myths and theologies to convey the rhythmic impact of the "Christ Force" to the current cultures and epochs. In line with this, a myth has been born within the Western culture that tells of the "Second Coming of Christ." The myth relates that something as special as the three-year phase of Jesus Christ's public works—which only a few people lived through with him—will happen again. The myth of the "Second Coming" can be traced to a piece of text at the beginning of the New Testament's Acts of the Apostles. They are the words of an angel in human form. After the disciples of Jesus Christ have watched their master's ascent to the heavenly heights, he speaks to them as follows: "Ye men of Galilee, why stand ye gazing up into heaven? This same Jesus, which is taken up from you into heaven, shall so come in like manner as ye have seen him go into heaven."

This statement does not mean that the Coming of Christ is one single event. Rather, it points to "the cosmic star's" next approach to humanity, the star that I call the Christ Force.

When I wrote this, my inner voice reacted most powerfully. I felt as if struck by lightning. For a long moment everything became black before my eyes. There was something like a spiritual slap in the face. What could that have been for?

My mind immediately had an answer ready and chided me for having landed myself in an unsupportable contradiction: How could I talk only of a "Second Coming of Christ," which is accomplished from the outside, out of the breadths of the universe—for then my efforts to effect an exchange with the inner child would become worthless. Had I forgotten the exercise in which we receive the Christ Child from the hands of the divine Mother, so that it may live on within us by turning our inwardness upside down? Here we have an exercise through which we experience the true coming of Christ. Only in this way can the experience have any personal value and successfully take root in the innermost heart.

Hey, stop! There is no contradiction here. The inner, personal experience takes precedence. However, it could not take place if a cosmic impulse did not lead the way. Such is also the case in the exercise described, where the inner child, that is to say the Higher Self, is first received from the hands of the Goddess—as it were from outside—in order to be internalized through the process of change. Thus there is no separation between inner and outer in this exercise, but instead there is mutual fertilization.

With this thought I was prompted to light a candle and give myself over to the inner Silence. I was hoping to learn what my inner voice wanted to tell me. I was reminded that in the year 2001 I had already had an indication of what to expect in regard to "Christ's Second Coming." Basically, it was revealed to me what road the "Second Coming" could take to bypass the mind's total control.

And here lies the main problem: Control by the mind—whether it be on the individual or global plane—has become so all-embracing that even a mighty impulse comparable to the "Second Coming" must find a way around. Today's consciousness is completely unaware of this amazing and pitiful fact.

The message that a hidden way to the "Second Coming" is to be found within the soul landscape of every single human being was intimated to me through a strange dream I had 18 days after the catastrophe of 9/11. That was on September 29, which is the Feast Day of the Archangel Michael.

Once again, the dream's beginning was set near our family home, which was an indication that it basically concerned some personal process of my own. In it, I am standing in front of my house and am mightily amazed to see a black-habited monk approaching. In a friendly fashion he asks whether I am willing to geomantically examine a parcel of land on which his monastery is proposing to build. My surprise is all the greater because the monastery has never before shown the least interest in my geomantic work. As I am weighing the matter, I catch myself thinking that I know something about this monastery.

In reality, you can search far and wide in our countryside and find no trace of any monastery. And yet during the three decades that we have been living here, I have known intuitively that a Christian monastery exists in the immediate neighborhood of our house. To be more precise, I have been aware of the parallel reality of a monastic family in our proximity and at the same time have wanted to know nothing about it.

Returning to the dream, my wife and I have been getting ready to visit my parents in a town about a hundred kilometers distant. But you see, the monk exclaims, he is also intending to travel to the same city. We feel obliged to invite him to climb into our silvery-colored car and travel with us. The mysterious monk occupies the seat directly behind me throughout the lengthy journey. This circumstance and the somewhat exhilarating sense that the two of us have a close relationship makes me suspect that the whole affair concerns a hidden aspect of myself. It is this notion that causes me to turn around suddenly and ask him whether he knows my book *Christ Power and the Earth Goddess*. To my renewed amazement he replies in a very convincing manner that he not only knows the title but has also read the book and found it interesting.

Looking back, I can see that this dialogue gives me the conclusive key to the dream's meaning. It confirms that the monk is a representative of that cosmic power that I equate with the newly appearing presence of the Christ. With this realization comes the shocking sense that the black-clad monk sitting behind me could be my own "double." In my dream, the premonition of these upsetting connections, which at that time were still not understood, causes me to turn round again and again and put further questions to the monk. In doing so I get dreadfully annoyed because he keeps trying to turn the conversation back to geomantic research on the monastic lands. This is not a theme that I find at all inspiring at the moment. Soon we reach the city limits. I turn around once more and abruptly tell the monk that he can get out here and make his own way. To cover my discourtesy I begin to talk some nonsense about the importance of our forthcoming visit to my parents.

At this point I should mention that there seems no direct connection between the theme of the dream and the processes that let loose the catastrophe of 9/11. This disconnect is quite a feature of the present sequence of Earth changes. A distinct chapter of the change sequence is introduced and its first steps completed, then a new theme is broached or an earlier chapter taken up again. Meanwhile the previously initiated process continues further on its own level and is drawn back into a later phase of the sequence for additional development. One must learn to follow this zigzag course or lose oneself again and again in the labyrinth of change.

This also happened to me in this case. In the following months certain events brought me closer to the meaning of the dream. After a talk I gave in northern Germany, for instance, a young woman came up and thanked me in a most touching manner. Apparently, she had been very ill and over several months I had quite often appeared to her and in a loving way led her on the road to health. I was astounded and told her that I had no inkling that I might have appeared anywhere as a healer.

Another example concerned an elderly couple who were unknown to me and lived at least a thousand kilometers distant. They wrote to me that at about 4:30 p.m. on a certain day I had appeared to them and asked them to make available to me information on the world of elemental beings. They did in fact possess this information.

It was not by chance that such surprises happened to me just during the few months following the dream described above—and thereafter not at all. There seemed to be an intentional delay in imparting the dream's meaning, which led me to think that it relates to the conscious evolution of a specific aspect of the human being, which—whether we like it or not—leads its own life. In writing on the subject, I have thought it best to talk of the human double or soul twin. However, Theosophy uses another expression for this aspect of ourselves and speaks of the causal body of the human being, which incorporates the cosmic wisdom and which we carry with us on our road through incarnation.

However, here we are not looking merely at a hidden dimension of consciousness. This much is clear from my dream. Rather, this concerns the renewed recognition of an essential part of the human being that has been excluded from our wholeness during our modern phase of evolution and, like a shadow, dwells "behind the back" of our lives.

In the dream, the monastery represents this dwelling that is situated "behind one's own back." This means that it is about that precise part of the human being that should be included in our relationship to divinity and the wholeness of life, a relationship that should constantly be renewing itself. The attribute "cosmic" is appropriate.

The dream carries a clear message that the separated "monk" within us is no longer content to be shut out. The cycle of Earth changes has come to the point where he is beginning to show a lively interest in the world situation as it currently unfolds. But that's not all. In the dream, this monk who has been repressed from memory forces himself upon my wife and me for the car journey to my *birthplace*. Perhaps he even would like to *incarnate*.

This is why I prefer to talk of a cosmic double. It is not just a matter of our personal strengths and qualities that would like to be

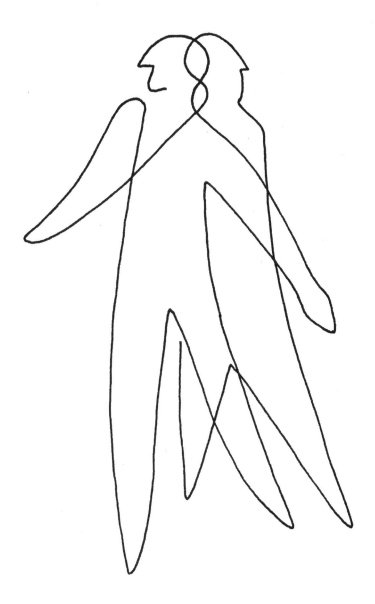

*A human being and his cosmic double.*

integrated into our waking consciousness; rather it concerns a part of our existence that could not incarnate at birth because the normal structure of the human being did not allow it. Among children who have been born during the last few years, matters may already run quite differently. However, as regards the monk in my dream, the birth process means he must remain behind in life's invisible realm, the etheric realm; he must lead his life on the etheric plane, parallel to his physical double.

When I speak of the "Second Coming of Christ," I am referring to the absolutely fantastic possibility that the original power of the universe, which in the West we call the Christ, could take form once again through the cosmic double's incarnating process. This would not happen through the customary way of incarnation but by a readiness on our part to allow "the Christ behind our back" to take a co-creative part in our change processes and in the new configuration of the being of humanity.

To understand exactly what I mean by the expressions "cosmic double," "twin souls" or "Christ behind our back," you can use the following exercise and pay attention to the quality and power that are made manifest thereby:

- Seek out a calm and peaceful place to go deep into the inner silence. Imagine that a person similar to you is sitting behind your back—indeed, you are sitting back to back.
- Now, let this cosmic double glide like a breath through your body, so that it appears in front of you. Look it in the eye and at first simply admit its presence into your Holon.
- While you are sitting opposite each other thus, both of you form a rounded space with its center midmost between you. Simply by imagining it, you build this space on the etheric, and on that plane it is real.
- Now, in your imagination, lead both of your figures simultaneously through this middle center. This will cause the etheric space to somersault. Allow the presence of this upside-

## After the Catastrophe: Humanity's New Inner Organization / 77

*You let your cosmic double glide through your body, and you look him in the eye.*

down space to spread itself out as widely as possible, and get a real sense of its quality.
- Try to incarnate this quality as deeply within you as possible.

You should make yourself so intensively familiar with this exercise that you have the feeling that your Holon is essentially enlarged. One is here looking at a quantum leap that results in one's normal Holon, which we discussed in the first chapter of this book, becoming a degree higher, deeper and wider. This can be a breathtaking experience. Can the body we are given indeed cope with this greater range? Must we not die in order to experience the powerful expansion that is here offered us?

**The Missing Energy Channels**
Happily, we do not need to die in order to realize these additional dimensions of our being presented here. The human vital-energetic body holds in readiness all the potential required for our accomplishment. We are equipped with etheric energy systems that lie fallow in the invisible realm for only as long as the time is not ripe to use them, and now the powerful process of the changing Earth is forcing them to show themselves. Like most of us, however, I too would know nothing of these hidden dimensions of the human body if I had not been quite roughly awakened to awareness of them by a weird dream that came to me on November 30, 2001.

In the dream, a stranger invites me to accompany him on a journey of discovery. Soon we are traveling along a country road in his SUV. The only excitement is a feeling of great expectation. We reach a narrow valley that winds between high mountains. The road leads alongside the steep slopes and it suddenly becomes so narrow that only half our vehicle will fit on it. The driver is not in the least disturbed but carries on without a care. From my angle of vision, I can see that only the two right-hand wheels of the SUV are running on the road. The two on the left-hand side are simply turning in the air. And yet, strangely, our road speed is quite unaffected.

Looking back, the signs are not hard to interpret. Quite clearly, they inform me that the driver is no ordinary man but a messenger from the spiritual world. The extraordinary power with which the two right-hand wheels have dug into the soft earth of the slope makes me think that this is a Master of the Earth Spirit, coming from the world of elemental beings.

In my dream we reach the opening to a narrow gorge that is both steep and deep. I expect the driver to stop here and let us descend to investigate the gorge on foot, but he does not do this. Although no path is visible, we drive as fast as ever into the dim opening. We are simply following along an old river bed, long ago dried up, which once must have enlivened the gorge with its loudly rushing waters. Our vehicle is an open jeep so I can look around freely. It is borne in upon me that the mighty gorge really represents a channel hewn in stone. I ask myself which of our past cultures was so advanced that it could bring such a gigantic work to completion.

Later researches into the dream's content have shown me that the work came about through an extraordinary cooperation between nature and the culture of the time. The constitution of the human energy body was such that our subtle body held an energy channel with latent potential. There were some old cultures that knew of this and by spiritual schooling and initiation rituals managed to activate these energy channels, using them consciously for the further evolution of humanity.

Back in the dream, my initial astonishment switches to a feeling of anxiety. First, I am afraid that our way may be barred by blocks of stone lying in the river bed. But nothing of the sort happens. However, instead of letting myself absorb the driver's attitude of inner assurance and calm, I am immediately overcome by another fear. Suppose that somewhere in the river's upper levels a sluice gate opens and we suddenly find ourselves drowning in floodwater?

The terrors that overcame me are testimony that knowledge of the mysterious energy channel was made taboo during the later epochs of human evolution. Obviously, people were punished, perhaps horribly, if they dared to use the potential power of the forbidden channel for themselves. There was the source of my fear—in the collective subconscious. It is certainly possible that an association of the elite may have used the leverage of the forbidden energy channel to consolidate their earthly supremacy and at the same time prevent their fellow citizens from gaining access to it.

Still dreaming, I am considering these shadow aspects of the newly discovered energy channel when we reach the end of our road. The car comes to an abrupt halt. Before us stands the chiseled stonework of the entrance gate to the ancient channel. The entry itself is barricaded by an iron grille. There is no way to get through. Shall we be forced to turn round and go back the way we came? While brooding over this, I peep through the iron grille and am disappointed to find that it leads into a banal suburban road. The hour is late in the evening and the road is poorly lit. There is little traffic—just a few pedestrians. A man goes by on an old bicycle.

The dream reveals that the newly discovered energy channel does not represent an esoteric secret, as I had thought at first, but instead has to do with the power and transfiguration of our daily life. The aura of secrecy that envelops it is merely based on the past, when the fact of its existence was banned from human consciousness, and so forgotten.

It is true that just such an energy channel is already known to the human world. It runs vertically down the spinal column and joins the cosmic and the earthly planes within us. However, the events portrayed in the dream indicate that we should not meddle with this channel, but rather be concerned with a horizontal one.

To discover the physical location of the missing energy channel, I tried in a meditation to reproduce in my own body the way to the channel's gorge shown in the dream. I often do exercises of this sort, using my own bodily resonances to help unlock the hidden features of a particular landscape.

To begin, I lay myself down and sank into the inner silence. Then I let myself picture the landscape of the dream and hovered around it, keeping my attention focused on any possible resonances that my own body could sense. I was seeking for correspondences between points on the route portrayed in the dream and various physical zones.

It was soon clear to me that when the spiritual master had come to collect me for the ride in his car, he had made a connection with the top of my head. The crown chakra that is located there is an energy center and responsible for our communication with the spiritual world.

The first section of the way, when we were following a normal road, leads me in meditation over the surface of my body down to the solar plexus chakra.

Arriving then at the edge of the belly region, the dream signified an escalation of the motivating force: We are now driving on two wheels. As related to the body, the way leads down the belly past the sexual region and disappears between the legs.

At this point in my dream, I had thought that all was lost. However, my companion at the steering wheel was not in the least confused. In the exercise, correspondingly, I am led to the region behind my back. There, in the empty space behind the sacral bone, I find the place I am seeking, the point where we entered the mighty gorge.

At first I am disappointed. What can there be of interest in the empty space behind one's back? Is there some correspondence between this point and the entrance to the great gorge that I knew in the dream?

Yes, there is—provided you include your own cosmic double in your meditation.

Firstly, you should envision the spinal column as the central axis of the body. From this axis there extend two bodies, not one only. The space in front is taken up by the physically incarnate human body and that behind by the etheric body of your cosmic double. They lean against each other, back to back. The composition is reminiscent of Siamese twins. They rotate as a pair around the energy channel that runs up and down the spinal column.

The great gorge starts in the middle of the cosmic double's belly region. Its belly represents the storehouse of archetypal powers from which our life processes draw their driving force. This is the inexhaustible underground treasure house of which the myths and fairy tales of countless people and cultures tell. It is not to be sought in some remote enchanted castle, but lies infinitely close: behind one's own back. It is merely that for the last thousand years its existence has been wiped from memory and banished to the world of fairy tales.

The drive along the dried-up riverbed should correspond to the course of the energy channel we are seeking. As represented in the body, it runs from the middle of the double's belly region to a particular point between the navel and the sex of the physical body. In everyday life, that is where the gateway with the iron grille is located. From that point, the archetypal power of life should be able to pour out freely to enrich the creativity of every day. Then we could realize the heaven on Earth for which we long. However, the barrier represented by the iron grille does not allow it. Consequently, our life is nourished only by those particles of archetypal power that succeed in passing through the iron grille, and we humans are forced to eke out our lives in ways that are more or less limited. But a fundamental change is approaching.

Based on the meditation described above, I have devised an exercise to experience and activate the newly discovered lumbar energy channel. The main aim of this exercise is to dismantle the barriers that dam up the fullness of life within us and hinder it from pouring out freely. To achieve this purpose it uses the healing power of the breath and various colors. Further exercises are included in the Appendix.

- You may stand, sit or lie down. Find your inner silence and connect with the quality of the cosmos through your crown chakra.
- *First Inbreath*: Draw your first breath in from the middle of

## After the Catastrophe: Humanity's New Inner Organization / 83

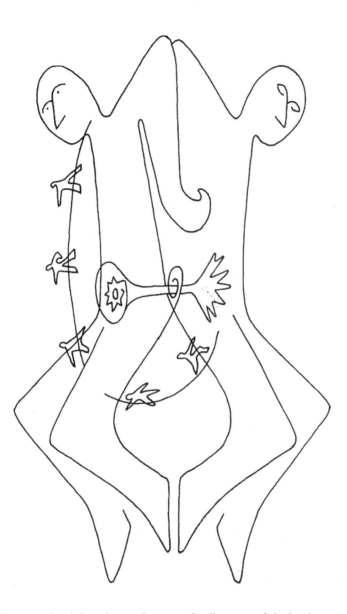

*The way that led me in my dream to the discovery of the lumbar energy channel.*

the universe; this breath is *white* in color. Draw it down through your body into the solar plexus chakra.
- *First Outbreath*: This outbreath is colored *golden yellow*. Let it stream down across the belly area, then between your legs and further up behind your back till it reaches the starting point of the lumbar energy channel (which is located in the middle of the belly region of the cosmic double).
- *The Second Inbreath* begins at this point. Its flow is now colored *green* to stimulate the regeneration of the lumbar energy channel. Let the flow of the inbreath stream through the channel from its beginning to its outflow point, which is located on the highest part of your physical body between the navel and sex.
- Now you breathe out, and the *Second Outbreath* takes on the color *violet* to effect the transformation of the above-mentioned barriers. Whirling in a myriad of vortices, the stream of violet breath passes throughout your Holon on its outward passage.
- *Third Inbreath*: Now the breath is drawn into your heart and there transformed into the perfection of the color *white*. This ensures that the power of the breath that was dispersed throughout the Holon is gathered together again.
- *Third Outbreath*: The assembled impulse is exhaled into the cosmos through the back of the head. There is no color anymore, but instead a crystal-like clarity that corresponds to the original space of eternity.

The rediscovery of the lumbar channel opens up a new dimension. We had a hint of this in the first chapter when we discussed the human Holon and the meaning of the relationship between the posterior and anterior spaces. To this horizontal relationship the discovery of the cosmic double adds a spiritual/soul dimension. This will gradually unfold through the renewed communication between the physical and etheric aspects of the human being, and in connection with the above-mentioned dimension of the second appearance of the Christ.

Now comes the lumbar channel. Here we are also dealing with a horizontal relationship, this time between the double's energy body, which remains in the etheric realm, and the incarnated human, who is fully incorporated in life. However, this relationship has more of a vital-energetic character. The expectation of a free exchange of the life force between the two poles of our being is inherent in the concept of the lumbar channel.

It is clear that the holistic consciousness of Earth has made good use of the deep rift that was created in the mental armor of human consciousness by the catastrophe of 9/11. This breach allows much that has been forgotten to be retrieved. A sequence of dreams that I received in December 2001 is testimony that there are treasures among them, in addition to the one already discussed.

The first dream in the sequence, occurring on December 7, makes a very clear statement. My wife and I are planning to travel to Monfalcone, an unattractive industrial city on the Adriatic coast near Trieste. I get the news that there is a boat anchored in our harbor that is going to Monfalcone. At once I run down to the harbor to ask the captain whether he will take us with him. The sailors make a remarkable impression on me. They explain that I have not heard aright. They will not be sailing to Monfalcone but to Montserrat—a Spanish pilgrim town that is renowned for its black Madonna. The sailors whisper to me that, apart from anything else, they could not take us on board because they have instructions to carry the King and Queen secretly to Montserrat. I feel confounded, because now our only course is to take the slow and tedious bus ride over bad roads with many changes.

The dream's message is expressed through its pairs of contrasts. The tedious journey overland along bumpy roads contrasts with the comfortable voyage by water. The industrial city of Monfalcone contrasts crassly with the pilgrim town of Montserrat. My wife and I contrast with the King and Queen.

With the help of the tension set up by the pairs of opposites I was made aware of a higher plane beyond the usual methods of communication. The symbol of the royal pair caused me to seek for this new plane in the realm of the crown chakra, i.e., in the head area.

The remaining dreams in the sequence only deepened the feelings of disappointment and discomfort that were aroused in me by the ending of the dream described above. They carried the serious warning that our present methods of communication will become useless in the course of the spreading Earth changes. To avoid total isolation and the resulting panic, people should already be making themselves familiar with a new possible way of communicating, using a forgotten channel that runs horizontally through the middle of the head, front to back.

It is certainly conceivable that there is really a channel of light between the Third Eye (which is the brow chakra) and the back of the head. At least two of the points of such a possible power channel are already quite widely known. The power center of the Third Eye makes holistic vision possible. The chakra at the back of the head is responsible for our relationship with the spiritual world and with the sphere of the dead. However, these two chakras do not represent the whole energy channel. Its tunnel reaches out further into the realm behind the head and includes the sphere of the cosmic double.

To help me further decode the message of the dream, I was given another dream on December 26, 2002, that showed me the wall of a house in section. In this view, the load-bearing portion is a solid stone wall that is standing in the background. To the front is the plaster layer of the façade. Between them, I can see piles of heavy building stone that serve no useful purpose. Suddenly the earth opens and engulfs the stones. The barrier they had formed, hindering the connection between these two parts of the wall, has vanished. The load-bearing wall and its façade now fuse together and lose their rigidity. They melt together to form a light wall that permits the transfer of communication.

The solid stone wall represents the double's etheric body, in particular the head area. Lying there are memories of our personal evolution over millions of years. This is a storehouse of cosmic memory that holds the experiences of past lives. Stored there too is the knowledge that we have gained through spiritual studies in the periods between incarnations. How helpful it would be in these times of transformation for each one of us to have access to the cosmic memory, to better direct us through the labyrinth of change!

The façade that appeared in my dream stands for our everyday consciousness that communicates with the outside world. The piled-up building blocks lying between the wall and façade symbolize the barrier that prevents us from entering the treasure house of our cosmic memory and joining the knowledge stored there to our waking consciousness. I am offering the following exercise to help you work on dismantling this barrier:

- You are present in your wholeness, well grounded and centered in the middle of your head.
- You change the physical form of your head into a sphere of light. From now on it will represent the Holon of your head.
- Imagine that you are taking the Holon of your head into your hands and lifting it very carefully off your neck. Very slowly you bring it forwards to the front of your chest and finally place it in your heart space.
- You let the sphere of light, which is the Holon of your head, rest in your midst until it is completely flooded with the power of your heart.
- Then let that sphere of light rise as gently as a soap bubble until once again it is one with the physical head.

As regards the changes within the human Holon, the situation at the end of the year 2001 can be described as follows: Instead of the vertical channel that runs between the root and crown chakras, there are two horizontal energy channels that have been reawakened and prepared to be bearers of human evolution. One of these is the lumbar channel, which, on the physical plane, is entered through the sexual organs. However, in this case the sole theme is neither the ability to reproduce nor sexual power in the traditional sense of the term. The spiritual and emotional planes are integrated within it, and thus we are dealing with the same power as actualizes artistic, playful and all manner of creative impulses. What is new and wonderful is that there is no longer any separation between above and below or profane and sacred. The lumbar channel encompasses the whole spectrum of the spiritual right through to the physical.

One can say much the same about the forehead channel, whose expression on the physical plane is the larynx. It guides the human ability to translate feelings, thoughts and intuitions into speech. However, in this case too, we are not merely dealing with the ability to communicate externally and take care of social relationships; just as important is the opportunity that the forehead channel provides to reach the depths of cosmic memory, and from there draw spiritual knowledge.

Another dream that I was given at Christmas 2001 best characterizes the new organization of the human Holon. In this dream I am first reminded of the legend of the seven heavens, and at once the seven heavens appear as realms of light piled on top of one another. Then I hear the solemn words, "The eighth heaven," and the eighth heaven begins to spread out above the vertical axis of the seventh heaven. It rises up like a tau-cross where the vertical arm is crowned by a horizontal one.

One can imagine that human energy systems may be made more horizontal, the vertical, hierarchical order thrown out in favor of several systems that are horizontally spread. These would be based on freedom of choice instead of hierarchical dependences, as in a rainbow where the whole spectrum of possible colors is arranged from one pole to the other along the horizontal axis.

In any case, this is not a matter where the old and new orders are joined in battle. One can experience a horizontal component even in the "old" energy channel that joins the seven chakras sequentially one below the other. There is a simple exercise for exploring this:

- Lie down and let yourself experience your vertical energy channel as it extends itself horizontally. When a person is lying down, the vertical will naturally become horizontal. Conversely, the lumbar and the forehead channels are brought into a vertical position. Now listen to the depths and the heights within you.

**The Heart Is the Midpoint**
The human Holon could not be transformed without the participation of the heart center. Just as one speaks of the lumbar and forehead channels, one can speak of a heart channel, with which would be associated with the following chakras:

- *The Heart Center in the middle of the cosmic double's body*
  Here we are looking at the root chakra of the heart channel. This is the region from which is drawn "the power of the first love" (using the words in the First Letter to the Churches in the Revelation of St. John) and which refers to the archetypal (divine) dimension of love.
- *The Heart Chakra of the seven recognized chakras*
  This is an individual generator of love's power, which, following holographic principles, is simultaneously a fractal of the heart of the universal Goddess.
- *The Heart Center in front of the breast*
  This heart center lies outside the physical body and is embedded in the emotional energy field of the human being. Its remit is to externalize the power of love that is generated within and bring it to expression in the world.

I should like to propose an exercise that enables one to experience the qualities of the heart channel propounded above. This exercise was inspired by a pine tree in Järna, Sweden, that had been brought down by a storm. It may have been the last message of its life. The pine tree was relatively young and grew on a broad rock, and its root growth, resembling a mandala, was therefore very much on the surface. After the wind had overthrown the tree, the magnificent mandala of its roots was standing upright:

- Imagine a root growth in the form of a mandala that stretches out from the middle of your cosmic double's heart. It contains anchored within it the quality of the first love, whose source is to be found in the midmost heart of your cosmic twin.
- The energy channel resembles the trunk of a tree lying on the ground and grows out horizontally from the middle of the root growth. Its crown is formed in your own midmost heart—a thick crown of leaves and branches that is shot with green throughout.
- After a little while a bud-bearing shoot begins to grow from the middle of the leafy crown. In the region of your breast a wonderful flower unfolds, taking the form of a mandala. Henceforward, its fragrance streams through all the areas of your life.

The organization of the new human being can be imagined as a cross with three horizontal arms. The vertical axis of the cross stands for the relationship between heaven and Earth within us, i.e., between the crown and root chakras. The three horizontal arms represent the three newly discovered energy channels: lumbar, heart and forehead.

This model of the cross with three horizontal arms was put to the test in the following spring at Capernaum, on the shores of the Sea of Galilee. At the end of March 2002 a group of Israelis and I had been busy with a weeklong healing project that focused mainly on the cityscape of Jerusalem. On the day before my departure, my work finished, I felt free to visit a place of my own choosing. I decided in favor of the ruins of the synagogue in Capernaum where the Gospels tell that Jesus taught and healed.

This was not due to my interest in biblical history. My decision to visit Capernaum grew from the experiences of the previous year when my daughter and colleague Ana Pogacnik-Meier and I had led a geomantic journey from Egypt across the desert of Sinai to the Sea of Galilee and ultimately to Jerusalem. We visited a number of sacred places during the days spent at the Sea of Galilee, and quite often I let myself dive deeply into the Holon of the Sea.

When I did so, my subtle perception always confirmed afresh that a powerful sphere of light is hovering over the Sea. My intuition whispered too that it is Christ's etheric presence that is working creatively out of this sun-like sphere. The powers and qualities set free during the time he spent teaching in Palestine are preserved and further cultivated in the memory of this sphere.

It is of course correct that the Christ Power—that is to say, the cosmic consciousness which in the West we call the Christ—has not been physically incarnate on Earth for the past 2,000 years. However, its influence has never completely vanished from the life streams of Earth and humanity. A worldwide system of such etherically interlinked light spheres enables the Christ Power—the Christ consciousness—to exercise a beneficent influence on the development of Earth and humanity even today.

Because the body of Earth and the body of a human being are holographically related, the system of these etheric spheres can be compared to a person's cosmic double. Just as the sphere-like centers of our immediate experience appear to be out of place—particularly in that they hover high above the Earth's surface—so the cosmic double is "hidden" behind our back.

On the morning of March 25, 2002, as I was standing on the shore of the Sea of Galilee, I decided to test my model of the new organization of the human energy body. To do so, I first built a heart relationship with the etheric sphere above the lake, to connect myself with the Christ presence that dwelt there. Then I imagined that the vertical energy channel together with the three horizontal arms of the above mentioned cross were present within my body.

There passed just a few moments of stillness, and then a powerful force began to work on me. First, the mandala of my heart chakra was stretched outwards by an invisible hand. A star-like image arose, with the heart chakra at its center. Its points reached to the crown chakra above and the root chakra below. They also stretched out correspondingly far sideways into the space of my Holon.

Next, the chakra in front of my breast was touched by the Christ power. An invisible hand stretched its "petals" outwards and finally bent them backwards. This backward bending accomplished something quite wonderful: The "petals" were fastened to the chakras at the ends of the forehead and lumbar channels to create a unified stream of power that joined the heart chakra to both energy channels.

The separation between the vertical energy channel and the three horizontal vital-energetic channels was demolished in a single moment. I found myself in the middle of a spherical stream of power that took the form of a figure eight or upright lemniscate (the symbol of eternity that is usually depicted as a figure eight lying on its side). The crossover point of the upright lemniscate was located precisely in the middle of my heart. After having the experience of this steam of power, I was gripped by the sense that the constellation of my chakras had been re-ordered in a totally new fashion.

Here is the corresponding exercise. It is a breathing exercise and best carried out in a standing position:

- Feel yourself anchored in your inner silence and rounded off all through the sphere of your Holon.
- *First Inbreath*: Draw this breath in from the middle of the universe vertically downwards to the midpoint of your lumbar channel.
- *First Outbreath*: When breathing out, push the breath back-

## After the Catastrophe: Humanity's New Inner Organization / 93

*The broadening of the heart channel, which I experienced when I was standing in Capernaum on the shores of the Sea of Galilee.*

wards and forwards through the lumbar channel till it reaches both chakras at the two ends of the channel.
- *Second Inbreath*: Draw this breath simultaneously from the two ends of the lumbar channel diagonally to the middle of your heart.
- *Second Outbreath*: Lead the outflowing stream of your breath diagonally to the two ends of the forehead channel simultaneously.
- *Third Inbreath*: Draw this breath from the chakras at the two ends of the forehead channel back by the same path to the heart center.
- *Third Outbreath*: Breathe the enriched breath out from your heart center into all the breadth of the life surrounding you.

Repeat the exercise, but draw the first breath from the middle of the Earth and not the universe. Consequently, the order of the breathing sequence is altered: You breathe first through the forehead channel. The lumbar channel follows with the second outbreath (see the illustration on page 250 in the Appendix).

Analogously to the fast changing structure of Earth's energy systems, one could say that we stand at the beginning of a process that will also fundamentally transform our individual energy systems. A brief description of its starting point will help you to comprehend this process. Because it is about to transform our whole chakra system, each of us should give sufficient time and attention to make ourselves clear about the matter. If you know enough about your own chakra system, it may be easier to understand the riddles of the "sickness phenomena" that can be associated with the change.

There are various ways in which to present the future chakra system. Let us take as our starting point the interpretation proposed by my daughter and colleague Ajra Miska and myself in our book *Schule der Geomantie*.

One aspect of the chakra system may be characterized as masculine (Yang), complementary to a second whose aspect is feminine

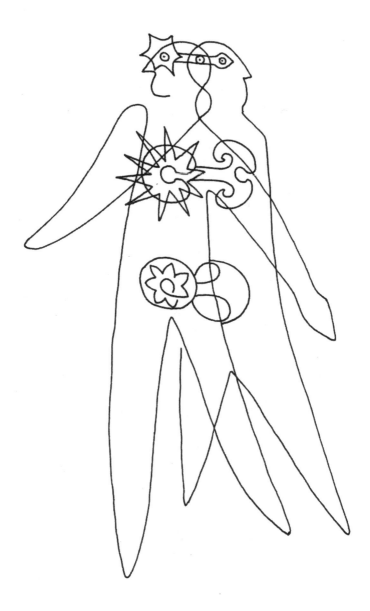

*The three horizontal light channels that have been newly awakened in human beings.*

(Yin). The Yang system is composed of the seven recognized chakras in their vertical sequence. In contrast, the Yin system has a circular formation. In all, there are eighteen little-known chakras, arranged in four circles around the heart center. Each of these circles resonates with one of the four elements.

The innermost circle—this one has two chakras located above the breast—is linked to the water element. The next circle—with chakras at the earlobes, shoulders and hips—is associated with the fire element. The chakras of the earth element are located at the elbows and knees, and there is one lying in the middle between the knees. Lastly there is the fourth circle, which represents the air element. The chakras belonging to it are located at the hands and feet; and there is one chakra that shines high above the head.

The vertical (Yang) system has the task of drawing the archetypal powers of heaven and Earth into our Holon. The circular (Yin) system, on the other hand, is connected to the powers of the manifested world. The heart chakra has the special role of acting as the limb that connects the two systems. For this reason the heart center is located in the exact middle of the vertical sequence of Yang chakras and also forms the midpoint of the circular system of Yin chakras.

The process of restructuring the chakra system does not diminish the central role of the heart center but rather strengthens it. This was brought home to me by my experience beside the Sea of Galilee. The heart chakra is expanded in the form of a mandala and thereby gains quite unimaginable power. The expansion of the chakra to front and rear contributes to this in the form of an energy channel. This channel allows the archetypal powers of the cosmic double to flow into the space and body of our Holon.

The danger that the new chakra system will take on too centralized a character is averted by two parallel energy channels. These are the forehead and lumbar channels, each of which is provided with its own horizontal chakra sequence.

We have to ask how, in the middle of a person's life, it is possible for one's energy structure to undergo such a complicated upset—and for most of us to be unconscious of it. A dream that I had on May 7, 2002, offers an explanation. In this dream I am with a group of people conducting a geomantic investigation of a large city. We have come to the end of the day and need tickets for the ride back to our hotel. I am asked to run to the nearest ticket window and take care of the matter. I am very surprised to see a seven-year-old boy standing behind the grille in charge of the selling. His hands quickly refashion each square-shaped ticket before he gives it to me, tearing each one with small, skillful fingers to make an eight-pointed star. I am amazed and ask whether these altered tickets are still valid. At that, he shows me one close up. To my astonishment, it has been torn so skillfully that no harm has been done to the control number, price and other printed information.

Given the background of the information we have about the archetypal image of the inner child, we can understand the dream's message to be that our own Higher Self is guiding the reshaping of our energy systems—and doing it so skillfully that the chakras and their energetic streams are functioning as usual. And yet, all the while, a star-shaped constellation of our energy systems is under construction.

Based on the eight-rayed star portrayed in the above dream, an exercise has been developed through which one can get a feeling for the changed shape of one's own energy system. In this, the model of the three energy channels layered one over the other has been replaced by the model of an eight-pointed star. Two horizontal rays or points represent the heart channel. The ends of the forehead and lumbar channels are joined to the heart center by four more rays. The vertical axis represents the relationship between Earth and heaven within the personal Holon (see the illustration on page 244 in the Appendix).

- At the beginning of the exercise, imagine the star with its eight rays as being within your own Holon. The star is posi-

tioned at right angles to the anterior surface of your body. Its rays reach almost to the edge of your Holon.
- Now the star begins to turn slowly around towards your back. You should take note of the sensations and feelings that accompany the slow turning of the star and be alert to what develops within you as a result.

CHAPTER FOUR

# February 2002: The Decisive Somersault in Etheric Space

## The Genetic Code of the Currently Approaching Somersault

THE READER MAY HAVE GOT the impression that the weeks following the catastrophe of 9/11 were devoted exclusively to the reorganization of our energetic being. That is not the case. In the middle of that autumn the seeds were sown for an event that shattered Earth and humanity, and actually occurred in February 2002.

You may well be asking, what event do I mean, for nothing noteworthy happened at the beginning of the year 2002. That we humans remember nothing is due to our success in letting the news of the worst destruction, which happened after the crime of 9/11, to sink into the collective unconscious.

On November 18, 2001 a dream let me know that a mighty upheaval was preparing as the Earth changes proceeded. At the time I was Hanover, busy with a city-healing project. The dream images transmitted such an upsetting premonition of cosmic overthrow as can hardly be put into words. However, I will make the attempt.

In the dream I see myself as one of a group of men who are moving a long, flat boat towards the ocean. The water reaches only to our knees, and we are exerting ourselves to bring the boat into deeper water. Two details strike me. One is that the direction in which we are maneuvering the boat is already prescribed for us. We can only push it along a narrow track that is bordered by two lines of stones that are quite strikingly tall. The other is my surprise at the unusual clarity of the water. One may even say that this is not really water but the etheric nature of water. It has the quality of liquid crystal. When we reach the edge of the much desired ocean, our efforts come to an abrupt halt. We gather excitedly at the edge and stare with astonishment into a gigantic hole: The ocean has vanished! Its mighty bed is empty.

Looking back at the dream, the gulf that opened before us then may best be compared with the gigantic holes in the landscape left behind by strip-mining for coal or copper.

As my dream-self looks into the hole more closely, I see that the dried-up ocean bed is alive with activity. Far below me people are diligently working. I take special note of some little children who appear no less active than the others. Canals are being built that resemble a widely extended network. But why canals? One can see far and wide across the dry ocean bed and there is no water there.

In answer to my question, my attention is directed to the edge of the dried-up ocean bed. We are standing on that same edge up to our knees in water, and I see that in many places this water above the hole has opened chinks in the dam wall that holds it back. Narrow gutters are forming through which the water from above is beginning to fill the gigantic ocean bed afresh. But I am surprised and shaken by the fact that the water now streaming down no longer has the etheric clarity of the water "from above," but is horribly dirty. The situation is becoming even more dramatic because the gutters are getting ever wider and it is possible that a flood of water may descend below. I worry what in this eventuality will happen to the children and adults below in the hole. They appear to be so deeply involved in their activities that they do not notice the danger of inundation.

Before I continue with the story of the dream, I should like to risk some attempt at interpretation: The dream is obviously set in

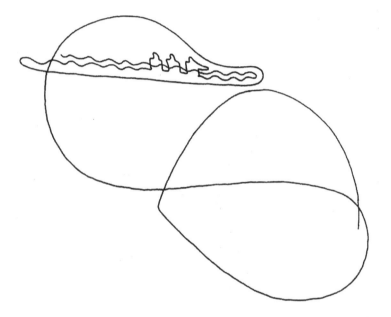

*The approaching reversal of Earth space: We stand at the edge of a gigantic hole and look down on the empty ocean bed.*

two areas that are glaringly separate. We, the main actors in the dream's story, are standing at the exact boundary between past and future when we look into the empty ocean bed. The "linear" track that so unambiguously prescribes the boat's course between the stones represents that segment of Earth changes that took place between the years 1998 and 2001.

That sequence of changes already past is characterized by a conflict: On the one hand, the boat is pushed straight ahead on a precisely prescribed track—therein I see the image of the old "linear" mind that is strongly oriented towards material things. On the other hand, I am surprised by the unusual clarity of the water through which we are moving—this image presents us with the new quality of etheric space that has manifested in the course of Earth changes over the last four years.

In fact, for the past few years humanity has been living "like an amphibian." On the one hand, we are forced to hold fast to the usual laws and patterns of the "old" materialistically oriented civilization. Its ways are prescribed by science and research that rest on the knowledge of matter. At the same time, we are unconsciously bathing in the refined frequencies of new qualities of etheric space that meantime have penetrated the invisible dimensions of reality. From within ourselves we are constantly receiving the call to leave behind the old patterns of thinking and behaving and open to the new multidimensional consciousness—which is Cosmic Love.

In the dream, when we reach the edge of the empty ocean, we get a glimpse of the next segment of Earth change. It is symbolized by a breathtakingly deep hole, which at first may seem disappointing. However, once this mighty hole has been filled with the crystal clear water that is shimmering "above" it and beneath our feet, a happier perspective results. The extremely constricted spiritual space of the (post)modern epoch is transformed into a deep and powerfully large ocean over which our consciousness can range freely. The dream's message, however, does not allow us to fantasize overmuch about future possibilities. Instead, two powerfully stated images make the future visible and bring it right into the present.

First, there is the image of the people below me who are so diligently working on the floor of the future ocean. These are the individuals from all over the world who are in tune to a greater or less er degree with the planetary change and trying in their own various ways to support the process. They are not waiting to see what the future may bring, but are already actively shaping it.

I saw them far below building a network of shallow canals, which I spontaneously compared with networking in the realm of consciousness. The voluntary initiatives that many of us are performing imply that we can already begin to work on the new paths of consciousness that will be created for the flood waves of water that in the future will fill the empty ocean bed. This will mitigate the impact of the flood, which might otherwise prove too destructive.

The mystery of why small children should be working there below us in the context of our future was clarified during the following winter. My colleague James Twyman from the USA posted a message on the Internet saying that there are children who are highly evolved spiritually and have been born in considerable numbers recently—the so-called indigo children—who are playing a decisive role in the actualization of the new etheric space. He has talked to four children in Bulgaria who, he says, have been cared for in a convent so they can work unhindered on the creation of an energetic grid that connects the indigo children with one another. According to them, the task of the grid is "to draw in the souls of other psychic children who would help shift the paradigm and to offer an energetic platform for the rest of us . . . to ascend to their level of awareness." Thus, the task of the indigo children is to help adults rise to the approaching quantum leap in consciousness.

The second relevant image has to do with the flooding of the dry ocean bed. I was shown that this has already happened. The dam that separated the living water from the mighty ocean hole has already burst. However, the water streaming down does not have the crystalline quality characterizing the water "from above," but is conspicuously dirty. In the further course of the dream this conundrum was explained as follows.

Our group is still standing on the narrow dam wall that separates Earth's future from its past. We are forced to wait until the dry ocean bed has been filled sufficiently for us to sail our boat upon it. I am finding it increasingly tedious to wait, all tightly pressed together, on the narrow dam wall of present time. My unpleasant mood is lifted by the discovery that there is a wooden table in our midst at which an old man is sitting. Sweets and desserts are laid out on the table, invitingly displayed for sale. They are made from dried, compressed fruit. I notice that they are contrived in remarkable shapes—step pyramids, for instance, and other archetypal forms. I find myself wondering what to buy as provisions for the forthcoming lengthy boat journey. Among the sweets displayed, I am especially attracted to a particular step pyramid, but at the same time I am afraid it could be poisonous. At this moment a clear feminine voice calls out my name: "Marco Polo Piccolo."

I am in fact called Marko. It is also relevant that in the dream I am about to embark on a long journey—like Marco Polo who traveled overland from Venice to China in the thirteenth century. Nonetheless, backtracking through past lives has shown me that I cannot give myself out as a reincarnation of my illustrious namesake—something my ego regrets. The nickname Piccolo, "the Little," has finally caused me to think about the suppressed shadow sides of my personality. Often enough my ego-consciousness has pushed them to the edge of forgetfulness and they are diminished, thereby shirking the—finally unavoidable—processes of change and purification. The precept not to be concerned with the past but to live in the present has also helped push unhealthy psychological patterns, personal dogma and blockages out of my field of consciousness. It is certainly true that they belong to the past, because that is when they had their source, whenever that was; however, the problem is that they create disturbances *in the present*.

The meaning of the peculiar name Marco Polo Piccolo is related to this problem, which also underlies the image of the old hunched-up salesman. The old man has sweets to offer shaped in archetypal forms reminiscent of by-gone cultures. Added to which, they are pressed together out of dried fruit; we are not looking at fresh fruit that has been bathing in the sunlight of our present day, but at preserved memories culled out of the abundance of past lives. And now I have no idea what to do with all this "sweetness."

Fortunately, on the following day I met a colleague, Wolfgang Schneider, who told me of his latest researches on the topic of stereotypes. He finds that repetitive dogma and rigid patterns are blocking the forward development of culture and humanity so decisively necessary at the present time. While the sequence of the approaching transformation and its associated overturning of space demand a high measure of elasticity and purity, the psychic burdens of the past can hinder us from freely following the change processes. Although we want the new with heart and soul, we are pushed back into old habits time and again. Our consciousness may already be at home in the broad spaces of universal thought—yet these burdens from the past hold us fast in emotional compulsions that repeat themselves over lengthy periods of time.

To physically experience how these ancient psychic burdens affect us, I let myself sink into contemplation of one of the dried-fruit step pyramids that play a role in my dream. I chose the pyramid I had thought of buying and which I believed might be poisonous.

I had scarcely got into the dream's images before I found myself in the midst of a wild and primitive landscape. Thunder roared and lightning shattered the sky. I was experiencing a much earlier civilization that saw as its task the taming of Nature's savagery here on Earth. In this, it was creating the right conditions on the physical plane for future cultures to unfold. I could feel how in the depths of my subconscious a hidden pattern resonated with this image. It was saying that Nature should be tamed and its wild strength cut down. Significantly, this pattern represents the exact opposite of all our present efforts to clear as many obstacles as possible from Nature's path and let her healing powers unfold freely once again.

### How the Reversal of Etheric Space Invaded Everyday Life

The dream about Marco Polo Piccolo made me realize that the war in Afghanistan was not the only consequence of the catastrophe of 9/11. The turmoil after the catastrophe caused a deep rent in the mental armor of humanity that was quietly used to sow a seed for the future overturning of Earth's etheric space. Isolated images from the dream help us form an idea of the approaching somersault.

To get a better understanding of the dream's message, I should like to return to the personal processes of spatial reversal that I discussed at the beginning of this book. It is noteworthy that the dream shows that the new living space, now gradually coming into being through the somersault process, is lying no higher than the shallow space through which humanity is presently moving. The "old" etheric space, in the form of the shallow sea with knee-high water, is located above; in contrast, the "new" etheric space is being prepared far below on the bed of the future ocean, and in my dream there is a great and vertiginous difference between the two planes.

We are here seeing the principle of reversal that Jesus cloaked in the words already quoted: "For many who are first will become last, and they will become one and the same." Like the Change itself, the principle of the Somersault is not a linear principle. Using the laws of logic, one cannot imagine how an old man can return to the vitality and purity of a "seven-day old child." The following exercise can help one experience the qualities that occur through the somersault process:

- Stretch your arms out on both sides and imagine that you are touching both sides of the space in front of you.
- Now very slowly move your arms horizontally towards each other to the point where they meet. The movement corresponds to the "old" linear space of my dream.
- Remain in this position for a moment during which the visible aspect of the place is symbolically made null. This position represents the dam wall of my dream, which separates the old spatial structure from the new.

- Then continue with the horizontal movement, crossing your arms more and more. "Soften" your glance and look inward through the "reversed" window formed by your crossed arms. This looking inward is mainly intended to give you a feeling for the reversed quality of space. You can succeed in doing this provided that through the whole process you make yourself one with your movements and with the space at which you are looking.

To experience the reversal of space in an exercise is one thing; it is quite different to feel it as a global reality in one's own body. We humans had this experience in the period between February 1 and 20, 2002.

I might never have discovered why, out of the clear blue sky, I endured such violent back pain over many days, if the fateful event had not previously announced itself in an astounding dream on January 31, 2002:

A pelican is born; it shines with unbelievable beauty. There is, however, a very serious problem. The pelican does not know how to feed itself. Many people are trying, one after another, to teach the luminous bird to feed itself, but without success. The failure of these attempts leads me to take the matter in hand myself. I remember, from the time in my life when I ran a farm at my house, about the dangers affecting newborn sheep. If the pretty-looking lambs do not learn to suck in timely fashion, their esophagus sticks together, and they die. Quickly I take the well-grown bird under my arm, intending to take it home and care for its feeding in my family circle. With the help of a sort of X-ray machine, I can see that its esophagus is almost completely stuck tight. It resembles a transparent, closed-up tube. This causes me to hurry even faster. To my great astonishment the bird itself comments on its serious situation, with the words: "It is already too late." These words are articulated in a quiet, calm voice with an unusually deep intonation. But a particular accent indicates that it is not a human that is speaking but another sort of being. My courage fails. Carrying the bird under my arm, I hurry even faster towards my house. Though it is still far off, I call to my wife to make a thin soup for the pelican.

My intuition tells me that the soup should be free of hard vegetable pieces so that the liquid can flow down the pelican's esophagus. It seems wise to heed the advice, but I feel as if crippled. In the event, nothing is done. I am just left with the hope that the soup will be cooked in the right way and that, despite the pelican's pessimistic statement, the feeding will have a successful outcome.

To understand the message of the dream, one must assume that the pelican represents the precious child born of the fearful catastrophe of 9/11. This figure of the newborn pelican symbolizes the crime's positive consequences, arising from the indescribable suffering occasioned by the sacrifice of so many. This was made possible despite the criminals' sick intentions because a seed of a health-bringing future was planted in their destructive deed through an act of mercy and love for humanity.

As described in the previous chapter, the embryo of a new kind of perfect environment developed from this seed during the months following the assault. If we pursue the parallel of a child's development in the mother's body, this means that in the night between the last day of January and the first of February 2002, the embryo's birth was approaching. This is the night that I dreamed of the newborn pelican.

Of course, it can be objected that the parallel of birth from a mother's body runs contrary to the facts. Birds are not born but hatched from an egg. How could a pelican have been *born* in that fateful night?

Nevertheless, the story holds true in spirit: In Western culture the pelican represents the brotherly love that Jesus taught. In the Middle Ages travelers returning from Africa reported that when starvation threatened its brood, the pelican would open its breast and gives its nestlings its own heart's blood to drink. Thus the pelican became a symbol of Christ.

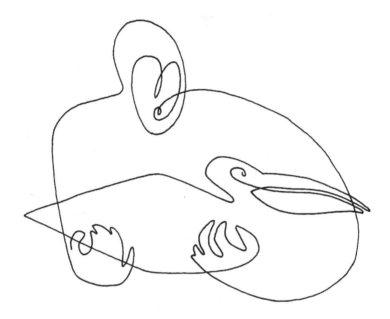

*Taking the newborn pelican in my arms, I run home and call to my wife to cook up a thin soup for it.*

When we look back on 9/11 and think of the people who sacrificed themselves to rescue their fellow humans from the danger of the "second death," thereby overcoming the contrary powers, they too stand revealed as such.

In this context one can understand the desperate concern, which so drove me in the dream, to act as quickly as possible to keep the new-born bird alive. The sacrifice of thousands of innocent people should in no way be brought to naught. Yet the bird's closed esophagus and its resigned words were testimony that this was a real danger.

When one looks at the circumstances surrounding the changes that will decide the future of Earth and humanity, it is all too easy to quickly lose hope. I am thinking of the world situation after the crime of 9/11. Instead of abandoning everything of lesser priority and devoting ourselves to the newly forming Earth, people's attention was diverted far away to the war in Afghanistan and fed ceaselessly with news of the possible dangers of further terrorist attacks.

The pelican's closed esophagus and his statement, "It is too late," refer to the danger that the wave of change produced by the catastrophe of 9/11 would subside, succumbing to the general human ignorance of the subtle processes at work within Earth and their own Holon. It was thanks to the sacrifice of the dead of 9/11 that this wave was released. Must humanity now go through something far worse before the walls of mental ignorance can fall?

We have not yet come to such a gloomy pass. The first dream sequence shows people who are concerned about the pelican, and we can see this as confirmation that there are human beings worldwide who hold the fate of Earth's kingdom very close to their hearts. However, in this there is a problem that is often overlooked. These are people who are certainly able to act and help. But for the most part they are locked in their own spiritual endeavors and stuck ostrich-like in various rigidly held ideas of what is important to the fate of the world. Unfortunately, such a state of consciousness lacks the sensitivity to feel the real necessities of the present moment. I have to admit that in this I am also critical of my own colleagues and friends on the spiritual path—and I cannot and will not exclude myself from the same criticism. The events of the dream show my panic-stricken reaction to the unexpected situation quite clearly. I only served as a figure that showed by example what one can do when it comes to space turning a somersault. Accordingly, my recommendation is:

- To let everything go immediately. Open yourself to the predicament and let it work on you so as to get a feeling of the seriousness of the situation. In the dream this relates to the state of my feelings when I decide to take over the care of the imperiled pelican.
- Do not let yourself focus on external reactions to the situation. Instead, take the event into your inner being and warm it in the glow of your heart. In the dream, the foregoing corresponds to my decision to take the pelican home and have a warm soup prepared for him.

The closing phase of the dream pointed to a serious problem of which I was obviously unconscious: I should have ensured that no hard vegetable bits were mixed in the soup for the new-born pelican. Yet the warning never passed my lips. How could the threat of danger leave me paralyzed, instead of impelling me to deal with it?

My unexpected inaction was occasioned by a problem I know all too well: my cowardice regarding the shadow side of being. My dream on November 18, 2002, when I was rightly called Marco Polo *Piccolo*, has made it clear that the epoch-making reversal of etheric space can be highly dangerous to life and limb—if the person is not prepared to work on the transformation of outmoded patterns and inner shadows.

The dream showed me the threatening outflow of dirty water—which means that there is a danger that the network of the new consciousness already under construction may be destroyed. Furthermore, there was the vision of the old salesman and the sweets he had on display made from dried fruits. The old salesman—the administrator of karma—tries to remind me as representative of modern man of the premier task in the Earth change process: to work unconditionally to release the powers that lie frozen in a person's inwardness because of karmic trauma or strongly held dogmatic patterns.

These are symbolized by the hard ingredients that if added to the soup would have hindered our attempt to open the pelican's sensitive esophagus to life's processes. Analogously, the subtle dimensions of multidimensional space, which have in the meantime been born, cannot unfold if the old dimension within us is not transformed.

We will return to this theme later. I would first like to relate my experience when I awoke in the night from the dream I have just described and went straight out into the natural world to see what was happening with the Earth.

There could be no doubt at all. I found that the spatial vibrations had reversed. What beforehand had had a positive quality now appeared as negative—not indeed in the judgmental sense of being "bad," only in the sense of a reversal of qualities. Externally, nothing appeared to be changed; the trees still soared skywards and the moon shone as always in the vault of heaven. The distinguishing mark of this new situation was that the physical form of space remained unaltered, but its etheric double had been turned on its head.

The dramatic impact on the environment of this reversal could be perceived the following morning in the mirror of the four elements. One sensed a powerful unrest. The water element reacted with threatening wave-like movements. It felt as space was rolling mightily like a ship. The fire element seemed to have withdrawn to the heights of heaven and formed a membrane of fire around the Earth. In contrast, the earth element, condensed to a ball, had obviously retreated into the Earth's depths. To my senses, the air element was acting in a way that made one's whole surroundings appear as if transparent.

To get a better grasp of these perceptions, we can draw upon the symbolism of my dream of November 1, 2001, which announced the coming somersault in etheric space. The experiences of the night of January 31–February 1, 2003 would correspond to that moment in the dream when I came with the boat and my companions to the edge of the ocean. Horrified, we gazed down into the gigantic hole; the ocean had vanished.

The relationship of the vanished ocean to the hole that replaced it corresponds to that episode in the night when I experienced the event as being "negative"; what had previously permeated the environment with all its fullness had suddenly disappeared.

One can reproduce the event in one's own body: Imagine that the etheric essence of your being (not your etheric body) takes the form of a figure of light and withdraws from your physical body. It moves out through the back of the head to reach the realm of the cosmic double behind your back, making a somersault-like movement so that the etheric essence of your being lands head downwards in your double's body. Now we have arrived at the position that I described above. Physically, you are still standing upright on your feet. Your etheric essence, however, is hanging head downwards—as if on the etheric plane you were hanging from heaven by your feet.

How does the exercise continue? It's clear one cannot remain forever in this conflicted position, where one's head is at one and the same time both above and below.

And yet space remained in this scarcely tolerable condition until February 20. It strained my spine so much that I could neither sit nor stand nor lie down. Others have told me of severe headaches; there were other people who were repeatedly sick. It might be a good idea to remember what happened to you during this time.

However, there were also people who enjoyed the month of February 2002. They partook of new ideas, intuitions and inspirations. We may understand this if we see that, for many people, the conflicted situation in their bodies helped them get free of habitual patterns. The "hole" created by the reversal was used as a gateway through which new impulses could enter.

Given this background, I was suddenly able to understand how Earth can put such spatial somersaults to use—the present reversal has not been the first and will not be the last. It is by such means that Earth can successfully release herself from the vise imposed by Mind, which has been forced on her by our present civilization. She is using the newly liberated space to take some fresh steps along the path of her intended change.

By February 20 the process had finally proceeded far enough to allow a glimmer of hope—but unfortunately this was linked to an unexpected deterioration in the situation. A dream the previous night had shown me a white cloud that hovered in space to the level of one's hips. I was confused. In certain places the cloud was thick, and from these there developed an intense white glow. I had never before seen a light that was so glaringly white. It could be assumed that the glow represented danger. In fact, in the dream several people suddenly appeared and tried to explain to me that the cloud originated from a chemical factory that had released a poisonous gas *in the past*. However, my intuition whispered that the source of the phenomenon did not stem from the past but from the future. If I would be patient, the truth would reveal itself.

On the day following the situation was at its worst. We had arrived at point zero. Nothing stirred; there was total standstill. How could life continue? Obviously, we had reached the turning point in the process of spatial somersault.

For a numerological point of view, February 20, 2002 was an absolutely unique day. If you write the date, European fashion, as 20.02.2002, it is a doubling of the year 2002. For this reason a worldwide meditation was called for two minutes after eight in the evening. This tripled the number of the year. The meditation took place on 20.02.2002 at 20.02 hours precisely.

Following the worldwide meditation I got the sense that an extremely faint impulse of general regeneration had come into being. Space had been totally emptied out and something had begun to fill it.

During my meditation the next morning I asked the Goddess to let me have some sense of the new spatial quality that had just come into being. Afterwards, a feeling of extreme ease and lightness flowed through me. My body dissolved into several layers. Each one hovered freely on its own in the air; the task of my consciousness was to hold them in harmony. I would describe the quality of this newly forming reality as extremely soft.

Three days later I had a very informative dream in this same connection. I am performing an exercise in the course of which I must bend deep down into a spring of water. The spring is located in a hole in the Earth. I am dressed in blue work clothes that have many pockets, large and small. All the pockets are packed full with various types of small tools. When I bend down over the spring, all the tools fall from my pockets simultaneously. I think to myself, what a shame that I have lent someone my toolbox. If I had had it with me now, I wouldn't be in this mess.

We should first look at the meaning of this last statement, because it helps us understand the current world situation. It means that if the tools of one's consciousness are packed in the toolbox of the mind, one will never notice the subtle quality of space after it makes its somersault. *Apparently* everything will be as it was before. It is quite otherwise if, consciously or unconsciously, one is vibrating in tune with the Earth change! In this case, one's feelings, thoughts and activities take on a completely new, yes, even reversed intellectual content.

In the same night my daughter Ana, more than a thousand kilometers distant, had a complementary dream. In the dream she is observing me in the downhill ski events at the Winter Olympics in Salt Lake City. She is overjoyed to see me on the point of winning the race. But her disappointment grows when I stop, standing some fifty meters from the finish line, and shortly afterwards start again on a slow zigzag course. Obviously, I will lose my advantage over the other competitors. Infuriated at my absurd behavior, she steps in herself, together with my wife. They successfully finish my run. Ana wins the gold medal, my wife the bronze. Then, for the first time, Ana realizes that when they stepped in for me, the two of them were not skiing in Salt Lake City. It was in Bled, Slovenia that they completed their victorious run, not in the USA.

She told me her dream over the telephone, and the impact of it shook me fully awake. It was sending a clear message that I was in danger of failing in the task assigned me in the present phase of space's dramatic somersault. Put plainly, it was obviously insufficient for me just to observe the situation while the spatial changes happened. As on 20.02.2002, when the worldwide meditation helped pull the change process from its threatened impasse, so too in the present phase some conscious intervention is necessary. The dream sends a message that I should not remain inactive, sunk in passive contemplation of the goal of Earth space's great somersault, and so miss the successful conclusion of the process.

Looking back, we can see that on that twenty-fifth day of February 2002 we were in fact at a point just three days prior to the end of the reversal process.

Fortunately, fate had so arranged matters that a day earlier nine women had come to Slovenia from various German-speaking countries to join me and take part in a seven-day geomantic seminar. Using the messages of the last two dreams, I presented Earth's serious situation to the seminar participants, how it was in the midst of a reversal process decisive for all our futures. I proposed that we should place our seminar at the disposal of Earth's consciousness, so that whatever it was now necessary to do or achieve could be accomplished. But in concrete terms, what did that mean?

Luckily, Ana's dream that I described above gave an indication. It is contained in the statement that Ana did not need to fly to the USA to step into my star-crossed ski contest. She was at home in Bled when she raced victoriously to the finish line.

Bled is a wonderfully holy place. A rocky island lies there in the middle of an egg-shaped lake. On it stands a pilgrims' church dedicated to the Virgin Mary. The dream made mention of the place to clue me that geomantic work on certain holy places can, in the decisive moment, release the right impulse to support the reversal of space.

Meeting with the nine women, I therefore proposed that in the course of the seminar we should work on sacred places for geomantic purposes and also for Earth healing. Thereupon we chose places within a radius of 50kilometers around my house. When we had visited these places one after another, they were finally linked together to make a network. This took place on the last day of February at a special spot, that—like a Jacob's ladder—has the ability to connect different planes and dimensions.

On the following morning we were to travel to Venice, Italy, to continue our seminar. Early, before we left, I ran far out into the countryside to test the quality of space. I was ecstatic. Etheric space was filled with a completely new quality that had a spiral-like twist to it. When my hand touched along this spiral, I had the feeling that it was composed of purest water.

For comparison, let us look back at the initial dream on November 18, 2001, which foretold the great somersault of etheric space that would soon occur. By the dream's inner clock, the moment has now come for the empty ocean bed to begin refilling with water. In fact, that has now happened.

This brings up several questions—for example, whether a small group of people does in general have the power to exert a creative influence on a worldwide process such as the reversal of Earth space. To this there are several complementary answers.

In the first place, one should look to the homeopathic principle that states that there is no limit to the spread of information in a power field of watery character. A group power field that is dedicated in pure love to the Earth cosmos is just such a watery field. Therefore, it is capable of limitless extension.

In the second place, one should keep in mind that Earth is an organism composed of an unimaginable number of planes. All the other planes cannot be equally affected by an event that is successfully completed on any one of them. It cannot be correct to expect a single quantum leap. One should rather imagine a number of such leaps, one following after the other, each of which will be relevant to the wholeness of the Earth. No one of them is final. Earth change is a monstrously complex process.

Another question is whether the approaching reversal of the planet's magnetic field, about which scientists have recently been talking, can be compared to the somersault in Earth's etheric space. My intuition tells me that we are looking here at two distinct processes, which however are coupled with one another by the principle of cause and effect.

The sequence of somersaults in etheric space represents an inner process of Earth change that will be accomplished on the vital-energetic, emotional and spiritual planes of the planet. It is through these that Earth is preparing herself to successfully perform the somersault on the physical plane as well. I feel certain that the threatened reversal of the poles of the Earth's magnetic field will occur much later—when the planet is inwardly prepared to carry it through in such manner that the richness of its life is not harmed thereby.

Now the moment has come to continue the exercise we began, and prepare for when Earth performs its somersault. You can help to achieve the same spatial reversal in your own inner self. In this case you are not only concerned with supporting the reversal processes of Earth, but also with creatively accompanying the same reversal in the realm of your personal Holon. Whatever happens in Earth space is inevitably transmitted to an individual's inner space through the resonance pertaining between macro- and microcosm.

- *Phase A*: Imagine that the etheric essence of your being (not your etheric body) takes the form of a figure of light and withdraws from your physical body. It moves out through the back of the head to reach the realm of the cosmic double behind your back, making a somersault-like movement so

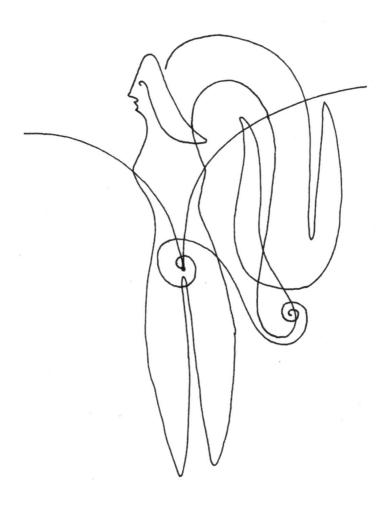

*One's personal space does a somersault—an exercise.*

that the etheric essence of your being lands head down in your double's body. Now you have reached the position that can be compared to the first phase of the spatial reversal. The physical being, as always, is still standing normally on its feet. Its etheric essence, however, is hanging head downwards, as if it were hanging down from heaven by its feet. Remain quite calmly in this position for a while.

- *Phase B*: Your being's etheric essence, which is hanging behind your back, now moves into the space of your hips. At the same time, spiral fashion, it folds itself inwards. It begins to turn inwards like a centrifuge, respectively forwards and back. The turning movements purify and at the same time diminish it until there is only a minuscule scrap still remaining in the hollow of your hips. Phase B represents the dramatic reversal in which—as described above—the life force of etheric space is almost lost.
- *Phase C*: Wait patiently. When you feel that the impulse towards a new spatial quality is arising anew from the nothingness, you should let it rise ceremoniously upwards through your body till it reaches the top of your head. Then, spread the new quality out through your entire Holon. Try to incorporate it in yourself as broadly as possible and bring it into your life. Phase C is to be compared with the renewed inundation of the empty ocean bed.

**A Renewed Earth Cosmos**

During the days following, the new, pleasantly watery quality of etheric space penetrated still further into the countryside. However, I could not understand why, in concert, the strength of the four elements should dwindle so rapidly. Rather, I had expected them to strengthen.

I went again and again to my observation points to discover whether the reversal, so successfully completed, had already had some positive effect on the quality of the elements. This was simply not the case.

On March 8 I made a note in my diary that it would obviously no longer be possible to judge the quality of etheric space by the condition of the four elements. I could not even perceive the water element on the surface of the Earth, but merely sensed its presence high in the heavens. In contrast, the fire element seemed still only present in the depths of the Earth. The quality of the earth element could not be detected at all. The air element seemed present everywhere. Where I delved further into my sense of it, I could perceive it as a general soul presence.

Still more catastrophic were the results of my observations on March 27. I could no longer make any meaningful contact with the four elements. I commented in my diary that the somersault of February 2002 had so utterly changed the Earth that the classical four elements might have lost all meaning in its new configuration.

Some dreams came to help me. There was one, for example, where a horribly weathered roadway has been covered with a smoothly perfect layer of asphalt. But people take no notice of the glaring difference. The newly asphalted road is once again strewn with gravel, just as was done before to patch up the fresh potholes.

The meaning of the dream could scarcely have been clearer. It pointed out that people are still traveling along the old road although the new one has already been built.

On Good Friday 2002 the mystery surrounding the fate of the four elements was finally solved. I had an important dream, which I forgot on waking. At first I was unhappy about that. Then I asked my inner self to lead me to a place through which I could approach the dream's meaning again.

I was not led to a single place but to seven, one after the other. Each place made an effort to bring me close to a specific content that was connected to the new "asphalt road." With hindsight, I can describe more precisely what each of the seven locations imparted to me. I was presented with a new and unquestionably cosmic configuration of the four elements. During the following Easter days, I went on repeated "pilgrimages" through these seven places to make myself more familiar with the new reality of the elements.

The first thing to note is that there is no longer any essential difference between the four terrestrial elements and the seven cosmic rays. Under the pressure of the forces that sprang from the epochal spatial somersault, they have grown together to make a synergetic whole.

Until now, the classical four elements—together with ether, the fifth element in the Western tradition—were the only tools that could work on earthly reality. The seven rays represented the complementary tools of the Universal Spirit, which were used to create the various dimensions of the universe. The elements and rays were essentially distinct from each other, because they belonged to different planes of being.

Now the two poles of creative power have grown together and synthesized. One can express it such that each one of the four elements has received a three-step cosmic superstructure (4 + 3 = 7). The three steps of the superstructure of the four traditional elements may be equated with the cosmic trinity in respect of the following qualities:

Wholeness and all-connectedness,
Divine-creative power,
Continued change.

By the second half of the year 2002 the cosmic quality of the four elements could already be "seen" by the inner eye. There was an especially strong brilliance perceptible in the aura of the energetic focus points, which had not been there before.

To better characterize the elements' new quality, one could also say that they no longer demonstrate the quasi-objective features of an energetic tool of creation. Rather they have become a soul-oriented tool of Earth consciousness. However, because the consciousness of Earth is an emotional consciousness, the four elements were given a strong emotional charge. These can be named respectively as follows:

*Water Element*: a young woman's presence, full of love and friendliness,
*Fire Element*: a deep and loving desire for perfection,
*Earth Element*: neutrality, mirth, heart's closeness,
*Air Element*: Transparency, breadth, curiosity.

Soon afterwards, "by chance" I was shown an exercise which you can use to perceive the four elements' new qualities in the natural world. I was on the train traveling from Muhldorf am Inn towards Salzburg. The carriage was almost empty. There was a woman sitting a few rows in front of me. Probably out of boredom, she stretched out both her hands above her back and then made the following gesture as if she wanted to show me it: she used her thumb to touch the middle joints, one after the other, of the other four fingers on the hand.

The gesture was familiar to me because I use a similar one when I want to identify the four classical elements in any particular location and attune to their current quality. My daughter Ajra Miskja had taught it to me several years earlier. To enter into resonance with a particular element, you make your thumb touch the *lowest* joint of the other four fingers on the same hand, to be precise:

Of the index finger for the water element,
Of the middle finger for the fire element,
Of the ring finger for the earth element,
Of the little finger for the air element.

The resonance bridge to the corresponding element is established by contact with the relevant finger. The other hand stays free to probe into the energy field of the place and key into its etheric space, feeling out the situation relative to the element in question.

My "railroad muse" had reprogrammed this exercise so that,

instead of the lowest finger joint, she touched the next highest. This brings you into resonance with the higher quality of the four elements—a quality that has come into existence because, since 20.02.2002, the "old" elements have taken on a cosmic superstructure.

**The Healing Centers**
The previous chapter reported on the cosmic double and the new organization of the human energy body. The present chapter is its complementary counterpart. I am attempting to record the results of the last reversal of etheric space as it made itself known on the macrocosmic plane of planet Earth. I am also comparing these new developments with the changes that have left their traces in the being of humanity, and looking for ways in which they correspond.

The epoch-making somersault in etheric space corresponds to the revelation that humans have a cosmic double. Just as a forgotten aspect of ourselves, "located behind our back," is activated by the course of change, so a spatial dimension in the landscape is also innervated. Until then this dimension had lain fallow in the spatial subconscious. My dream of November 18, 2001, represented it as the empty ocean bed.

The discovery of the lumbar channels in human beings corresponds to the cosmic superstructure that has been added to the four elements. Activation of the lumbar channels joined our sexual powers to the cosmic sources of creation. Our being's creative capacities in the realms of Earth and spirit were no longer to be separate. Also, since early in 2002 there has been a spiritual dimension indwelling in the four elements—the creative tools of Earth—which, although they are still active on Earth, has raised them to become tools of the universal spirit.

Now we must look at what corresponds to the forehead channel. I found this in the "healing centers" that I perceived for the first time in various cityscapes during the early part of 2002. They were not in evidence before the epochal somersault in etheric space. A dream that I was given on May 21, 2002, in Järna, Sweden, has helped me understand the special role of these centers.

I dream that my stepmother and her spoilt daughter have come unannounced to my home. They have settled themselves in the house without asking if I am happy with the arrangement. When I enter my home, they are engaged in a bitter quarrel. My stepmother is making a row in the kitchen while her daughter is lying barebacked and sulking on a bed. I am busy and several times I run past her extremely broad and naked back. Finally, I make a firm resolve I will plant a kiss on her broad back, if, the next time I pass by, she is still lying there. Then I see myself on the terrace in the act of taking off my briefs. As I take them off, I am wondering how it can possibly be that my daughter—not the stepdaughter—is already dead.

A dream of this sort resembles a complicated hieroglyph that can only be deciphered with much spiritual exertion. I spent many nights and meditations wrestling with the meaning of this dream before I could explain its message to my satisfaction. In this I was helped by experiences in my Earth healing work during the spring of 2002. At that time I was being impelled time and again to go to points in cityscapes where I found a power with an incredibly healing quality. In such places one can sense a special, love-filled freshness, as if heaven had bent down there to kiss the Earth. This experience helped me categorize the kiss that in the dream I had applied to my stepdaughter's broad back.

The young woman's broad, bare back represents the landscape. However, this is not a landscape of the free and natural world, but rather a landscape of the towns and cities, stressed by human beings. They carry the stamp of their inhabitants' egotism and mental turbulence, as the sulky stepdaughter demonstrated.

Now I would have preferred to see the *Earth Mother herself* and her daughter Natura in my home—not her estranged double! But the day-to-day reality of the landscape now presents only her alienated and mutilated face, and that is what the consciousness of Earth must show. To counter this, a geomantic system of healing centers was developed that can overcome the generalized blockages that work against the processes of change in these places. These blockages have come into being through the paralyzed force fields of the estranged landscape.

It is a fact that up till now I have only found these systems of healing centers in cities, where most of the deeply disturbed and chaotic places are to be found. The new centers usually lie in sequence along a particular axis. They display a full spectrum of different qualities and functions. In my dream the connections were represented by the remarkable image of me taking off my underpants so that I can think of my dead daughter.

This image symbolizes a whole spectrum of reality, beginning with the archetypal powers of Earth's deep places and stretching to the high planes of the spiritual world where the dead dwell. The systems of healing centers follow the same model. There are always points that have to do with the sources of archetypal power; some enable converse with the spiritual world and some uncover the whole realm between.

One example is the Hungarian capital of Budapest, where I had business in mid-April 2002. I had already worked with this city on the River Danube in the previous year. However, this was the first time that I could perceive the existence of a spiritual axis that runs through the city along the line of the Danube. Along this axis I found a sequenced row of energy centers that I identify as healing points. The axis draws its power from a ley line and runs beside it through the cityscape.

The axis begins in Tabán, an old part of the city that today has been almost entirely destroyed and replaced by a gigantic road junction. This is home to those sources of archetypal power whose task it is to nourish the city's vital-energetic system. The other end of the axis is located in the meditative environment of Margaret Island, which was the site of an important convent in the Middle Ages. This is a place that is connected to the spiritual world. The axis also runs through the showy Parliament Building, which guards the national treasure in its midst: the crown of St. Stephen.

All the points that reveal the quality of healing centers are ranged along the axis. Their activity enlivens the whole spectrum of the axis, including pouring out their message through the entire country, which comes about through the activity in parliament.

In a city landscape one usually finds healing centers in places where Nature is still to some degree alive, such as where parks are so laid out as to be connected. I found this latter case, for example, in Bremen and Regensburg, where the old ring-shaped ramparts have been transformed into parkland. From there, their creative activity spreads through the whole structure of the city.

One should emphasize, however, that healing centers are usually identical with energy focus points, and have always been so. The impulse of change is bringing them new life and penetrating them with the quality of healing. This enables them to work co-creatively to heal and reshape Earth's geomantic system.

I investigated a similar case in Quito, Ecuador, during the fall of 2002. This capital city of Ecuador lies in a relatively narrow valley between two ridges of the Andes mountain range. The healing centers are ranged in sequence along the two long sides of the cityscape, and especially grouped where they debouch into the open countryside. The centers are connected in a network, such that the threads of the network run diagonally through the structure of the city.

While at a workshop in Ljubljana, working on the system of healing centers in Slovenia's capital city, I have "seen" how the "kiss" of a healing center comes into being to heal an alienated landscape. I was shown, high above me in the heavens, a ribbon composed of healing energies. The ribbon is the source of the healing potential of all the rays that are anchored deep under the Earth. At the point where they touch the Earth's surface—or rather where they bore through it to reach the Earth's depths—a healing center comes into existence.

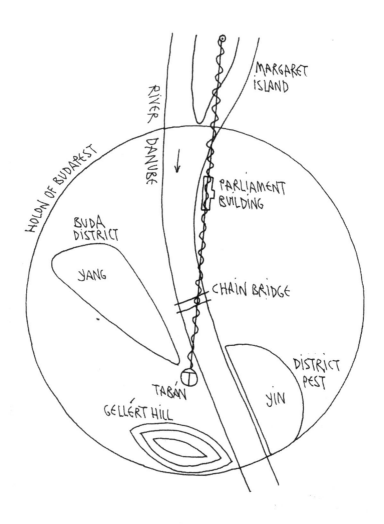

*The Holon of Budapest, showing the spiritual axis along which healing centers are beginning to manifest.*

CHAPTER FIVE

# After the Somersault: The Rushing Torrent of the Great Purification

### The Double's Shadow Aspect

ON APRIL 14, 2002, I had a dream that dampened my enthusiasm for the momentous consequences of the somersault in Earth space during the preceding February. For a whole month I managed to preserve my happy mood by ignoring the message contained in the dream. After that my defenses broke down.

The dream tells a very simple story. My publisher phones and asks me to come to his office at once. When I get there, it is to have a document from the Ministry of Culture thrust under my nose. It tells me that the Ministry is prohibiting the appearance of my new book because they themselves are on the point of putting out a similar book, and to publish two similar books at the same time makes no sense. I am deeply offended and very angry, and immediately I lodge my protest. In the first place, I say, mine is a modest book that is confined to a particular country, so there can be no question of competition. Furthermore, I believe the prohibition to have been instigated by the Minister himself. Since I know him personally, I have a good mind to storm into the Ministry, angry as I am, and rebuke him for a decision that is hindering my *highly important* work.

After I had awoken, it was clear to me at once that my protest had no merit and was no more than a confused mixture of exaggerated humility and puffed up grandiosity. Anyone who opens him- or herself to the message contained in the dream will see that it is an attempt to make me aware of a fault. Many qualities of which I can in no way be proud and with which I have to struggle continually in my daily life are precisely caricatured in order to open my eyes to specific truths.

Seen from the perspective of later events, the dream indicates that the present moment in the sequence of Earth changes may contain some other meaningful task than the one to which up till now I have been devoting my attention. It was a summons for me to let go of my firmly held preconceptions regarding the general direction that the present Earth changes were taking and open myself to the possibility of a change of course. The symbol of the book, prepared by the Ministry and differing from mine only slightly but still in a momentous way, pointed to an unexpected alteration in the process of change, about which I had not the slightest suspicion until then.

Thus far, the message can be read from the structure of the dream. However, I had no practical idea of how to act differently if I decided to follow the dream's instructions. To be more precise, my waking consciousness was striving—to put the matter politely—to fake things for my unconscious, because it feared stepping off into the unknown. This led to my carrying on with my life as if the fateful message had never been received.

This incorrect behavior towards the voice of my own soul reaped its revenge on my health. After three weeks I began to suffer from various physical complaints and I saw myself virtually forced to listen afresh to the forgotten message.

A dream came to help me. The first image showed me confined in my exterior consciousness, "exerting" myself unsuccessfully to answer to the pressing question of what was to change in my work. In the dream I am a passenger in a car packed full of luggage, which the driver is persistently trying to maneuver into too narrow a parking space. He steers it countless times in and out of the little gap.

Finally, this back and forth motion is too much for me. Angrily, I get out of the car, saying that I have had enough. Instead, I take myself to my room, but when I go in, I am totally surprised. Since I was last there, the right-hand side of the room has been rearranged and shines as never before with order and beauty. However, when I turn round and look at the left-hand side of the room, which was behind my back before, I nearly have a heart attack. Never have I seen such terrible disorder. Deeply shocked, I perceive that there are some beings there that are making themselves busy amid the disorder.

When you analyze the dream, you see two distinct levels. One is concerned with the exterior, egocentric consciousness and is symbolized by the car that is trying to park in a narrow space *in front of* the house. The other has to do with one's own room, which is on the house's first storey. This symbolizes the personal, spiritual-soul space in relation to the inner space of the human psyche.

The dream characterizes the modern man or woman as a split being whose consciousness exists in two spaces that are separated one from the other. On the one hand, you are trying in vain to find your parking place in the street of the exterior consciousness. On the other hand and separate from this effort, your psyche's inner life runs on a "higher" level of space. This recognized and often attractive fact is of itself quite tragic. It leads—and this is the heavy and starkly emotional meaning of the dream—to a further, disastrous division, which actually happens within a person's inner space, and of which modern men and women have scarcely an inkling.

The dream images suggest that we can imagine this startling division to be a hard and fast separation between the light and shadow aspects of our inner world. In the dream the two aspects are represented by the two different directions, left and right, to which I direct the eyes of my psyche.

Where I looked first, I saw the new and wonderful organization of the being of humanity, that recent progress on the path of change is gradually bringing into existence. This is being effected—as already described—primarily through the rediscovery of the cosmic double and the three horizontal energy channels that are balanced by the power of the heart.

However, when I turned around, it was to see the terrible disorder behind my back. I believe this to be identical with my personal shadow. What most upset me was that up till now I had considered the realm behind my back to belong exclusively to the blessed presence of the cosmic double. How could the space that until then I had seen as connected to Christ's second coming simultaneously resemble a dusty junk room? Could there be some further, unknown dimension dwelling within that much prized space behind a person's back?

For me, this was a burning question, and it was 11 days before I stumbled on the answer. I was given it in the final minutes before my departure from Venice in May 2002, when I was already sitting in the boat that was to take me to the Marco Polo airport, from there to fly to Sweden. I had been away from home for several days, working in Venice with an American group, but, oddly, the message did not reach me till the last moment.

The boat's captain let us know that it would be 20 minutes before we cast off. There would be time for a coffee on shore. The prospect of a cup of coffee did not motivate me so much as my interest in art. The boat, which traveled between the city and the airport, was moored opposite Giudecca Island and just in front of the Church of Santa Maria del Rosario, which is called the Gesuati. It had been at least 15 years since I had visited the church, because, when preparing my first book on Venice, I had been unable to fit it into the water city's geomantic system. I knew, however, that the Gesuati Church was home to some admirable paintings, and I decided to use the wait to enjoy its works of art.

I began my tour on the right-hand side of the church. The first painting impressed me very deeply. It was a masterpiece by Giambattista Tiepolo, painted in 1748, and shows the Virgin Mary appearing to three holy women—Rosa of Lima, Catherine of Siena and Agnes. The divine Virgin is portrayed above their heads, and in

# After the Somersault: The Rushing Torrent of the Great Purification / 133

*The upper portion of the painting by Tiepolo, showing the universal Goddess whose essence is mirrored by a little bird. She has handed her divine son to St. Rosa of Lima.*

front of her a tiny bird is sitting on an iron perch that seemingly bores through the painting's background. Mary's beauty is indescribable—she is rightly depicted as the universal Goddess.

Rosa of Lima is standing below Mary, holding the newborn Christ Child, whom Mary had given her to rock and cradle. Here let us remind ourselves of the archetype of the inner child to which we devoted the beginning of this book. It is really wonderful to see how St. Rosa incarnates the seeking human's love-filled attraction to the inner core, represented by the Christ Child.

To her right stands St. Catherine of Siena, whose right hand holds a crucifix that is nearly as tall as she. Quite remarkably, her attitude so skillfully reflects the suffering figure of Christ on the cross that one has the impression that the crucified one is actually to be found behind her back. The crucifix, held upright and colored rusty brown, is like a shadow.

One must ask what archetype St. Catherine can represent if it is not the archetype of the seeking human in relation to his or her cosmic double, whose archetypal figure is related to the Second Coming of Christ—behold the crucifix—and who dwells in the shadow realm at our back. Let us remind ourselves of the cosmic double's role, which we have suggested is to connect the appropriate person simultaneously to the sources of the archetypal power of Earth and the wisdom of the cosmic realm.

I had finally found an inspiring representation of the connection between the principle of the inner child and the archetype of the cosmic double—hence my enthusiasm over this painting.

Tiepolo's painting had more to offer, however. It also hinted at a shadow aspect of the cosmic double that until then had been unknown to me and that I had never even considered in connection with his exalted figure. But here the painting portrays the shadow not only through the rust-brown figure of the crucified redeemer, but through St. Catherine, who seems to be bearing the figure of the crucified one "behind her back" and also is wearing an instrument of torture, the crown of thorns. There is yet more. Lying on the ground

## After the Somersault: The Rushing Torrent of the Great Purification / 135

*The lower portion of the painting by Tiepolo, showing the shadow as well as the light-filled aspect of the appearance of the Christ.*

is a skull that her lovely right foot is touching. In a quite amazing manner the symbols of the shadow are shown to be the attributes of her own person.

Returning to the dream in which I described the two contrasting halves of "my room," the image of St. Rosa of Lima with the little inner child in her arms represents the newly ordered half, which blazed with light. St. Catherine, bearing the symbols of suffering and approaching death, represents the shadow side, which struck me with such unbelievable fear when I turned round and saw the muddle and disorder behind me.

During my flight to Sweden I pondered the message of the painting in the Gesuati Church. One might say that the human cosmic double had taken on the role reflected by the archetype of the crucified Son of God—a role that one does not usually associate with the lofty ideal of the spiritual master. Was the painting saying that the heavy burden of personal sin and other wrongs against the wholeness of life has been taken away from humanity and placed on the shoulders of our cosmic double? What would be the meaning of such an act of grace?

The answer is clear. It would mean that no human being would have to suffer being totally blocked on his or her personal path. Since our shadow would be taken and offered up in full sacrifice by our cosmic double, we would be free to work on our further evolution without collapsing under the burden of past guilt.

Until then I had struggled mightily against accepting the symbol of the crucified one as a symbol of the Christ. Even in my book *Christ Power and the Earth Goddess* I could find no honorable place for the suffering of God's Son on the cross. I could not accept that another might take over one's personal problems and, using some external authority, raise them into the light in an act of redemption. In my view this idea lacked the sense of personal responsibility that enables us to grow inwardly through experiences gained by struggling with our personal shadow.

Now I was confronted with an image of the human microcosm

in which the principle of redemption is joined with the principle of full responsibility for one's personal spiritual path. Here, we do not have to wait for some future redemption, but when we work on the transformation of our own shadow realm, we can experience the redemptive power in present time as being one of the indwelling energies of our own Holon.

Yet this is in no way the same thing as self-redemption! The redemptive power springs from humanity's cosmic twin, which I have already described as an aspect of the Christ. It is related to the personal "I," and yet simultaneously preserves the all-embracing quality of a universal being.

When I arrived in Stockholm, Sweden, I wanted to go on immediately to the coast at Solvik, opposite the island of Mörkö, which was the place where two years earlier I had experienced the nearness of death. It was there, whispered my intuition, that I had first known the terror-struck feeling that I experienced afresh in my dream about the two halves of my room.

At that earlier time I had been visiting the coast of Solvik near Järna for one of my seminars, to look for sites where one could perceive elemental beings. I had found a place by the sea where I perceived an etheric entryway into an unknown dimension. I paid little attention to the phenomenon at the time, going on to busy myself with the task in hand. It was in the hours and night following that the devastating consequences of this brief encounter showed themselves. I fell sick, had terrifying experiences of the near approach of death and a dream that confirmed for me that I had encountered a power capable of shearing through the threads of my life.

Later I was able to integrate this experience into the composition I wrote of my journey through the different aspects of the Divine Feminine in humanity, for I realized that I had encountered the black aspect of the Goddess. But the real drama of this encounter remained hidden from me until my experience with the painting in the Gesuati Church. It was not until after that visit that I became aware that two years previously in Solvik I had first looked my shadow in the eye, an ordeal that till then the grace of the cosmic double "behind my back" had kept hidden from me. It is only when a person is ready, spiritually and emotionally, to confront the fearful truth of the shadow that this sight is granted one.

It is absolutely impossible for the exterior consciousness to imagine the horror of one's little faults, so harmless in one's own mind, when they appear projected onto the white screen of cosmic perfection. The mere sight of them can be shattering, right down to the foundations of the soul. One is therefore always protected from this experience by grace of one's cosmic twin, who takes the burden off one's shoulders. Voluntarily, he allows himself to be "crucified" in place of the person involved.

However, the essentials of the situation have been altered by the Earth changes that the cosmos has initiated. We are being called upon to rediscover and live up to our lost wholeness. It would be impossible for us humans to think of swinging ourselves up to a higher evolutionary level without the old burdens being previously transformed. One by one, they must be lifted off the shoulders of our cosmic twin and, with his loving help, transformed in the light.

Returning to my dream and the document that my publisher had received "from the Ministry," this insight solved the mystery of its message. It was that I should temporarily abandon my concentration on changes in the phenomena of landscape and people and instead devote myself primarily to the present phase of Earth change—one that involves liberation from one's personal shadow. As soon as I understood this to be the meaning of the two dreams described at the beginning of this chapter, I was given an appropriate exercise. The ease with which this came about is an indication of how relatively simple it has now become to handle these ancient burdens. The task is unavoidable, but from the side of the cosmic consciousness—so the angelic world tells us—all possible tools will be made available to ease us through it.

The task will not be achieved by an exclusive appeal to divine grace. For karmic reasons, one is obliged to make one's own contribution, even if it is minimal. When one has taken the first steps towards changing one's personal shadow—for example, by performing the exercise proposed below—one receives help in such full measure that the transformation process is relatively easy.

The new exercise was first performed by about fifty seminar participants from throughout Sweden, at the exact spot opposite the island of Mörkö where two years before, confronted by my unredeemed shadow, I had come close to death. Those attempting the exercise should know that it is very powerful and you may need to recuperate afterwards. This is particularly true if it has been a while since you last performed such spiritual work. It is best performed on a day when you have nothing else scheduled and do not plan to drive a car or operate machinery for at least four hours afterwards. Despite or perhaps because of these precautions, the exercise is very much to be recommended for the freedom and clarity imparted; it helps realize the Christ Within, ensuring smoother adjustment to the changing Earth. The exercise is as follows:

- Ground yourself and visualize your Holon's protective cloak around you. You are connected and centered.
- You are conscious of your cosmic double's presence behind your back. Like a forgotten twin, your double hovers behind you, back to back.
- You check whether he is burdened with the heavy load of one of your unreclaimed shadow aspects. This may be an aspect already known to you, one with which you are consciously dealing. However, you can also expect to be surprised, and should be open to whatever is ready to change.
- You let the selected shadow of your double glide through your body so completely that every physical cell is touched.
- Then bring the shadow forward till it hovers in front of you and you recognize its "face." Look closely at it, or rather, feel out its presence. Allow its powers to unfold freely before you, however hateful they may prove to be.

- When you sense that these powers have reached their maximum extension, ask for the grace of change.
- After that, bring the shadow, head down and facing forwards, into the vortex of change. There will be a succession of somersaults like the path through a centrifuge, a sequence with many reversals. You can also allow the color violet to flow into the process.
- At the appropriate moment, the concentration on the vortex relaxes somewhat. Now watch what arises from the vortex *at this moment*. It may be a being of light, a light-mandala, a beneficent feeling or much else. The power that was previously frozen in the shadow is now a positive quality and free, and it steps forward.
- Take this power into your heart and let it become part of you. If you have been working to change an estranged fragment of the Earth's soul, the redeemed power will spread out over the world. Give thanks.

When all the seminar participants had performed the exercise individually, they gathered as a group in a calm and sunny place to report on their experiences. I was burningly interested in this exchange. For one thing, I was unsure whether my method for the transformation of the personal shadow would work as successfully for other people as it did for me. For another, I had an uneasy feeling that the change process was running almost too smoothly in some cases. This was in stark contrast to the previous year, when there had been all too strong a sense of a leaden and powerful opposition.

The participants' reports were reassuring, sometimes even inspiring. Some told of serious trauma experienced in their younger days that had surprisingly emerged in the course of the exercise and could be successfully transformed in the light. Others reported amazing insights into onerous past lives, whose trauma they had successfully approached and given over to the vortex of change. There were people who had been stuck in various forms of obsessive compulsion and reported that their causal origins had been successfully brought out in the change process.

These happy results of the exercise to change the personal shadow encouraged me to a further step. On the following day—May 20, 2002—I proposed that the group should contribute to changing the collective shadow of humanity. According to holographic principles, a group that has consciously dedicated itself to love and truth can act as the representative of humanity and work on changing collective shadow aspects.

Yet the question still remained—what, in general, did we understand by the *collective shadow?*

In this connection, I am highly suspicious of any talk of a "devil." And yet, as regards the collective shadow, one can imagine that countless personal faults, and conscious malevolence even more, can accumulate bit by bit to become a giant cloud of negative power, the mega-shadow of humanity.

But a further step is necessary for complete understanding. During the long epoch when humanity was becoming alienated from itself—the epoch that, using a concept culled from archeology, I have earlier called the Age of Iron—there were individuals and groups who time and again consciously tried to misuse the mega-shadow to strengthen their own claims to power. This repeated communication, with its consciously negative value, gradually imparted a maleficent intelligence that would finally enable "him" to break free of his masters. A relatively self-sufficient contrary power has been created, whose exclusive interest is to survive and grow in importance.

Because this frightening intelligence sprang from the split consciousness of modern men and women, the shadow being has also split into two opposing "giants." One of them clings tightly to matter and encourages people to seek after material power. The other is oriented towards theology and philosophy and strenuously tries to lead people astray to embrace false ideals and illusions.

These two poles of the mega-shadow have already been mentioned in the dream sequence in Chapter Two, where they fought and inflicted deadly wounds on each other in connection with the catastrophe of 9/11. At the beginning of the twentieth century, Rudolf Steiner gave the name of the God Ahriman to the pole that is drawn to matter. This was the name of the Persian God of Darkness. His counterpart in the Light was called Lucifer. The meaning attached to this latter name tells of a power that leads us too far into the light, so that we drown in illusions.

These considerations helped the Järna seminar group to clarify the concept of the shadow aspect of humanity. We then looked for the most appropriate place where we could celebrate the ritual for changing the mega-shadow.

This ritual follows a pattern similar to the one for changing the personal shadow. A circle is formed and dedicated to the divine light and the love of the heart. An aspect of the collective shadow that has been discussed and chosen beforehand is then manifested outside the circle, i.e., "behind its back." Afterwards, the mega-shadow is drawn through the power ring of the circle to its center, there to be perceived in its true shape—not drawn into the center of the circle through a person's body, which in this case would be too intense an experience. Next the shadow being is delivered over to the vortex of transformation. At the end of the process the force, which is now changed and positive, is given back to the appropriate country.

Again and again during the following months I asked seminar groups in different lands to work with me on transforming those aspects of the collective shadow that had decisive importance for their particular country. In the course of these exercises people went through experiences so deeply upsetting as to be scarcely describable. Yet at the same time the colors and light emitted by the liberated shadows were sometimes of special beauty and subtlety.

When I returned to my house, I decided to follow up on the message contained in the dream in more logical fashion. I had been expressly desired, in the name of the "Ministry," to temporarily put my personal projects on hold and devote myself to the collective task of transforming the mega-shadow. I invited two of my colleagues to assist in carrying out this goal. Together we brainstormed various plans for changing the shadow aspects of peoples and the world, and help spread these changes over all the Earth.

Eventually, we proposed that on a Sunday evening in June 2002 at nine o'clock, Middle European time, everyone taking part in the project should connect with all the others worldwide and, in the silence, work on changing the shadow. We offered three different methods. Wolfgang Schneider included the angelic world. Peter Frank recommended a meditation to help recognize the personal shadow. My own plan is detailed above. It is hard to imagine the magnitude of the healing power that is accumulated when people come together from all over the world, not once but several times, to devote themselves to the transformation of the world's shadow.

### The Cunning of the Contrary Powers

Thanks to the activities described, I had finally found the way to achieve the task suggested to me by the dream about "Marco Polo Piccolo," recounted in the previous chapter. You may remember that it was then, with the help of certain images, that I had been instructed to regard the transformation of the human shadow aspects as a priority task. Without such a commitment there could be no essential progress towards realization of the new world structure.

Although this gave grounds for celebration, I was far from being satisfied. The symptoms of my illness had not gone away but taken another form. To this was added another alarming dream. Obviously, in my struggle with the dark power I had overlooked something important. Otherwise I would not have received the following warning during the night immediately after our first exercise on transforming the personal shadow.

This is the story of my dream on May 21, 2002. Without my being aware, I—a pacifist—have been so expertly manipulated by a group of youths that I find myself involved in a violent struggle with the police. The situation has escalated to the point that I have no option but to flee with the youths into an attic storey and there hide myself. This, however, does not pass unnoticed, and the police are right on our heels. The attic has several windows but only one cramped opening through which a person can crawl in and out. The youths press a wooden board into my hands and then retreat to the furthest corner of the room. I am supposed to hold the board in front of the opening so that the police cannot force their way in. When the police realize that we have taken refuge behind the board, they begin to fire at it with heavy volleys. Fortunately the board is made out of sawdust and a number of thin wooden sheets glued together, so that the bullets, though very many, cannot penetrate it. I am taking the situation in deadly earnest, firmly holding onto the board, and thereby making myself an accomplice of the youths' behavior. They, however, are not taking the situation at all seriously but are grinning and even appear to be making themselves merry over my efforts. Under the constant fire, my board begins to disintegrate at its edges, and in the end we must surrender to the police. As soon as we have crept out of our hiding hole and gathered in the street, the whole affair with the police is resolved. They vanish without bothering themselves about whatever misdeeds we intended. At that moment I notice that water is streaming into the street from all around me. A terrible flood is coming.

Without trying to interpret all the aspects of the dream, one can affirm that it basically means that I am the victim of a conspiracy. Obviously I have been infiltrated by a conviction or ideology that has robbed my deeds of the power of truthfulness. In consequence I have been drawn into particular decisions and dealings that make no sense in the eyes of the Whole. The dream's last image would signify that by reason of failings still unknown my powers have declined and are being swept away, which is crassly contrary to the urgent demand to gather all available resources and direct them in support of Earth's transformation.

As I searched for the dead corner of my perception to which I had fallen victim, there came back to memory all the dreams to which I had given insufficient attention previously—for example, the dream of the wounded eagle resembling a hen that had settled on my head. I had dreamed of it nine days following the crime of 9/11, after I had asked whether I would run into any danger now if I flew to the USA. Its meaning then was that the power that had been working against the epochal change of Earth and humanity had had its back broken. It emphasized, however, that another danger threatened, of whose nature I was unaware.

In its crippled state the eagle had developed a strategy to survive and heal its deadly wounds. For me, this was demonstrated by its having sunk its talons so deeply into my scalp that on waking I could still feel their sharp points in my head.

The eagle's firm grip on my head could be interpreted to mean that the dark power's future survival strategy lay in itself taking ownership of the model that future-oriented movements had developed to support the life processes of Earth and human changes, and thereby care for the whole future of life. The contrary powers are also interested in an assured future, although they work against the further evolution of Earth and humanity. Unfortunately, this paradoxical condition has in the meantime become our reality.

The political tumult and agitation that was led by the USA and followed worldwide after the crime of 9/11 also served to camouflage the cunning maneuver in which the contrary powers are trying to board the ship that is lying ready, raise the anchor and sail away into the New Age.

At this point we should clarify more precisely what is to be understood by the terms "contrary powers" or "dark power." These phenomena draw their might from unclarity, and we can therefore allow ourselves no ambiguities, once we have decided to bring the shadow into the light.

In the first place, we should reject any part of the old dualistic model of the eternal battle between good and evil. Those who hold fast to this model must also hazard that they—although apparently standing firmly on the side of good—will sooner or later find themselves on the other side of the artificial line separating them.

In the second place, I must make clear that when speaking of these cunning contrary powers, I am in no way referring to the cosmic powers of darkness that Hildegard of Bingen, so celebrated today, has named the "Cherubim of Darkness." By "Cherubim of Darkness" she had in mind the hidden workings of contrary powers on the cosmic level. Their task is to mercilessly attack all developments in the universe that, knowingly or unknowingly, contain seeds that would destroy the cosmic balance and harmony. This means that the beings or cultures affected must either adjust to the dark power or suffer the painful consequences of its assault, which ensures that over time any unhealthy developments will be corrected.

In my book *Earth Changes, Human Destiny* I recounted in some detail one of my early experiences in the present process of Earth change, which affirmed that in general the process was initiated so that thereby the Earth could receive a gift of divine grace. The "Cherubim of Darkness" were drawn out of their body—and thus also out of the body of humanity. Expressed in the picture language of the Apocalypse, the "red dragon" is bound in chains for the span of a thousand years (Rev. 20:2).

Thus the cunning maneuver referred to above cannot be the work of those contrary powers that have their role within the cosmic Holon. It can only be carried through by the dark power that works within the planetary Holon. In the chapter discussing 9/11, this power was described as a shadowy embodiment of the two extreme poles of alienated humanity, which either lead people astray or force them to submit. The personal shadow aspects of which we spoke at the beginning of this chapter duplicate these two extremes, but work within an individual's personal Holon.

Mention has already been made of the fact, unfortunately too often buried, that the contrary powers on the planetary level incorporate intelligence, perhaps through having absorbed the evil plans of some of the people who have trafficked with them. In the end, this has enabled them to construct for themselves an individuated demonic consciousness. I am not using the word "demonic" in the sense of a value judgment. It is rather the case that this consciousness, unnaturally amassed from disparate parts, possesses no indwelling kernel of the divine but only an empty will to survive.

Beyond the divided forces of power-hungry materialism and degenerate idealism there is yet a third malignant power. It hides itself behind the manifestations of the two contrary forces and uses them for its own survival. As need arises it takes on the masks of Ahriman and Lucifer, either to rule humanity by reason of our own greed for power or to lead us astray through perverted presentations of belief.

However, it would be inappropriate to draw too dramatic a picture of the demon of human evolution. Rather, my rare experiences of it suggest that we are dealing here with an evil-minded little man who nurtures a compressed form of cunning and humanly developed intelligence. The dream image of the grinning youths, amused at having enticed me into a serious fight with the police, represents the demon quite fittingly. A coarse and malicious glee in other people's troubles goes hand in hand with *absolute* lovelessness. A cold intelligence hides the absence of compassion.

It is just as wrong to demonize and banish this revolting being as it is to accept it. Instead of wavering between these two extremes we should realize that the demon is concerned with the collective shadow aspect of humanity. Since each one of us represents a part of humanity, it is also concerned with our own shadow aspect, which we cannot escape.

We may ask whether it would it not be sufficient, as proposed above, to devote ourselves intensively to transforming our own shadow aspect and thereby redeem our own portion of the collective demon.

Indeed, work on the transformation of our personal shadows is much to be recommended. But it is not enough, because we know that the demon's intelligence has developed an artful trick by which, *of its own accord*, it can withdraw itself from any attempt to redeem it. This is the problem with which we must currently deal.

On November 29, 2002, in Laibach, Carinthia, I dreamt a dream that illustrated the alarming state into which the new strategy devised by the contrary powers had brought the change process. The dream was as follows. There is an unknown power that wants to treat my throat with acupuncture. It will use a thick steel needle at least 18 cm. long (that's over 7 inches) to bore into my pharynx. My fear and outrage were such that that I awoke from sleep.

The dream continued after I had gone back to sleep. Now, a former general is conducting me through landscapes that have been devastated by long use for troop exercises. The army has left them in a very pitiable condition. At the end of the tour the general takes his leave and goes away. When he is already far off, my inner child, whom I am holding in my arms, begins to cry after him despairingly, asking him to come back. Hearing the sound of the general's name, I am reminded of the name of my spiritual master. My ego refuses to take the hint. I am displeased that the child should be shouting for the general and I make every effort to quiet the child.

I might have understood the meaning of the dream immediately, had I not been convinced that the forces that worked *consciously* against the changes in Earth and humanity had, in 1997, already dropped out of the process. There certainly remained those two extremes of the power that we humans had mutilated, but I thought that after causing each other such mutually fatal wounds they would have come under control.

If I had not been blinded by such long held ideas, I might have been able to see that the first part of the dream demonstrated the discrepancy between my expectations and the actual state of affairs. I had expected that the contrary powers would be ready, after the withdrawal of the Cherubim of Darkness and their mutual disaster of 9/11, to renounce their claims to world domination and place no more obstacles in the path of Earth healing.

Quite opposite to my expectations, the dream meant that the dark power had used the unfavorable world situation to develop a new survival strategy. That highly dangerous needle that was to operate on my throat was actually an indication that an aggressive strategy was preparing that would attack the human creative potential. The larynx chakra stands for the ability to do spiritual work.

In the second part of the dream the spiritual master appeared in person to remind me of a simple fact: Because of the worldwide destruction perpetrated by people and cultures—all instigated by the maleficent intelligence of the contrary power—the motivating force behind it cannot be expected to disappear without going through a process of deep transformation. This was a wake-up call to my supposedly responsible consciousness, which at that time was unfortunately affected by the above mentioned blindness and not capable of such insights.

A second dream joined in to enlighten me. My bladder, which represents my emotional world and resembles a giant pumpkin, is lightly cooked and opened on one side while still in my body. A maidenly being slides her upper body into the slimy, transparent mass. For her, the healing warmth of my belly is a limitless delight. I am even glad to be carrying a child in my belly until I notice that the being's eyes have suddenly taken on a demonic sheen. Her face becomes grotesque.

A few nights later came a complementary dream with a further warning. In this dream I am undergoing an operation on my throat to remove a sort of polyp. I am suspicious because, during the course of the operation, the surgeon has changed into an unpleasant being that is devouring the mass of dried blood that has peeled off from the excised polyp.

With hindsight, we can see that the last two dreams make fairly clear the type of action that the demon of human evolution has chosen to ensure his survival through the epochal Earth changes. He is finding persons of a frivolous or reckless disposition who are unconsciously prepared to let him drive his energy into their feeling world and there let it bathe in their feelings of self-absorption. By this I mean feelings of self-pity, self-satisfaction, self-guilt, self-glorification and so on.

Once the alien, demonic energy has made its nest in the feeling world of an "innocent" person, it begins to feed on the creative energies that are produced in his or her larynx chakra. In particular, those qualities are attacked that make up the personal identity—hence the dream's allusion to blood. The resultant weakening of the organism's defenses allows that warm bed in the belly to be re-occupied.

But is there anything new here, and how is it related to Earth's epochal changes? The dark powers have always taken every opportunity to build themselves nests in the feeling and creative worlds of humanity, there to nurture negative feelings and destructive thoughts. That is simply part of their mission, to hold up a mirror to those who have chosen to follow a false trail. I ask why alarming dreams should be constantly drawing my attention to the matter.

The answers to these questions came to me on January 10, 2003, when the threat of war with Iraq decided me to take some action to promote peace. On the very night following, there came a dream to help me find a way, and with it the inspiration for formulating the meditation that later would be spread worldwide. This is to be found in the Appendix on page 259. The dream also contained an important message for our struggle with the dark intelligence.

This is the content of the dream. With numerous companions I am ready to leave harbor and set out on a long journey. Suddenly, from deep within the belly of the giant ship comes the sound of two explosions, one after the other. The noise is audible but muffled. The ship lists dangerously, first to one side, then to the other, and then rocks a few times to and fro. There follows a third and much more violent explosion, whose sound however is again muffled. This time the ship takes on an extremely heavy list, so that the water laps the very edge of the deck. I stop breathing. But we overcome the danger. The entire crew storms on deck to see what is happening and find the cause of the assault. Surprisingly, not far from the ship there is a group of grown men who are standing up to their knees in water. They are holding each other by the hand and stand in a circle. Half of them are clothed in startlingly white laboratory coats, the other half in pitch-black suits like tuxedos. I am utterly shocked. It is not at all becoming for a band of men to be clothed in such a manner and hold each other's hands as if for a spiritual ritual. Then it happens. The circle turns towards me and the men focus on me in particular, not letting go of each other's hands. They look at me with an evil grin, as if saying, "See what we have done." I will never forget that deceitful, cruel and mocking look.

At first I was struck by the parallel with my dream on September 22, 2001, when I had watched the two giants who were engaged in a life and death struggle and, just like the men in the group, were standing up to their knees in water. They represented the polarities of the traditional consciousness that were smashed to pieces by the catastrophe of 9/11.

In this new dream the first two explosions bring to mind the two planes that crashed into the twin towers of the World Trade Center. That was when the two extremes of the contrary power collided with each other.

At the dream's third explosion, black and white are standing together in a circle, shoulder to shoulder. Have these powers, until now so hostile towards each other, finally learnt to put aside their black and white thinking and relinquish their utter opposition to each other?

That cruel look, so happy at my unhappiness, that was cast at me by these former opponents, now allies, left no doubt in my mind that their mutual understanding was feigned. I realize that I am witness to a cunning trick by the demon of human evolution, who is trying to imitate the road to incarnation that was taken by the Christ Power, so as to appear to people quite renewed.

In our earlier discussions of the cosmic double as the twin soul of humanity, I mentioned that in order to appear quite renewed among us, the divine inspiration that in the West we call the Christ took an unexpected route. This route runs between the consciousness and the feeling world of all who have learned to make a bridge between the opposites within themselves, and maintain their inner peace. The harmony of opposites and the preservation of inner peace together form a recognition code.

This is why the last dream image dismayed me. I was worried that it pointed to a new danger, that the path to redemption may be blocked and perverted to a survival strategy of the dark power. It was through the two qualities that I have just described as a recognition code that the path of redemption was protected. That circle of one-time enemies who are now allies is an indication that they plan to fake them.

In answer to my worry, on one of the nights following I was given a glimpse into the technique that the demon of humanity is using to encroach on the spiritual-soul's protected path. The starting points are our most common customs, which we unconsciously follow to ensure our continued subsistence here. I am speaking of the pieces of silver that we must pay to the alienated world structure in order to continue to be counted among its sub-tenants, and relatedly of our beliefs, which we must betray in order to be better understood by our fellow men and women.

The gulf between what a person is in truth, and what he or she shows outwardly and pretends to be, can from now on be fatal. The discrepancy, often scarcely noticeable, can be used as an entry port by the dark power to make itself a nest in the feeling world of the unsuspecting person. What most people see only as harmless indecisiveness can lead to a person becoming an unwilling accomplice of the dark power that is trying to hide itself from the approaching epochal purification.

I do not wish to spread fear and uncertainty. It is clear to me, moreover, that we live in a world structure that desires us to submit ourselves to institutions and social norms. But this is not a matter for concern; external things do not possess the potency to be exploited by the contrary power. What I have in mind are the often unnoticed, inward acts whereby one denies one's true Self. It is they that can attract those fate-filled consequences. Every one of us is summoned to use our intuition or develop the necessary sensibility to recognize cases of this sort in timely fashion and deliver their causes over to the transformation process.

The following explanatory dream was given me on February 12, 2003. I am standing on the edge of a footbridge holding a naked child "seven days old" in my arms and looking into the depths of the sea. Suddenly an unknown voice within me says that I can throw the child into the sea without him incurring any danger, because the principle of buoyancy will bring him to the surface again. I find this enlightening and, untroubled, throw the child into the deep water. For a long while there is no sign of him, and fear rises up in me. There is no joy when he does surface, because waves are constantly washing over his head. Apart from this, there is the danger that he may be hurled onto submerged rocks. I stand there, pondering whether to leap into sea and save the child. As if I were a prisoner of some unknown power, I cannot make the decision, although the need is clearly urgent.

I have made available two different sorts of tools so that we may successfully avoid the danger of becoming an accomplice of the contrary power that is striving to withdraw itself from the great purification: the first of these is a clear ethical attitude, and the second, clear emotions.

## A Clear Ethical Stance

The right ethical basis is contained in the seven letters to the seven churches in Asia, which are to be found at the beginning of the Revelation of St. John. In my book *Earth Changes, Human Destiny* I suggested what they mean for the future and the evolution of humanity. They are encoded in a special way and contain precise instructions as to one's inward disposition during the current epochal change and expected turbulence. Also, there are seven references to ethical conduct that offer us unparalleled help in our struggle with the demonic power. They are linked to the seven cities of the Apocalypse where there were early Christian communities:

*Ephesus: Love!*
At every moment, let yourself follow the call of your heart. Test yourself whether, in any given situation, you really embody the voice of original love.

*Smyrna: Do not fear!*
Never shy away from whatever your personal or collective fate may be sending you. In every situation preserve your inner peace.

*Pergamos: Change yourself!*
Always be ready to pursue the incessant stream of change. Test out for yourself which aspect of you or your creations is next in line to ask to be changed.

*Thyatira: Be truthful!*
Test yourself to be sure whether in any given moment you are not hiding some aspect of the truth either from yourself or from others. Always investigate your heart and your spirit to ensure that you have not become the victim of self-deception.

*Sardis: Be whole!*
Always remain emotionally aware of your many-layered wholeness. Keep the great round of your being embraced in your consciousness and anchored in your mid-point.

*Philadelphia: Be loyal!*
Do not forget who you are and to what ideals you have inwardly pledged your faith. Keep reminding yourself afresh of your spiritual calling.

*Laodicea: Be decisive!*
In every situation you are given various possibilities from which to choose. You are called upon to make decisions. The one thing that you may not do in this epoch of great change is to remain undecided.

## Clear Emotions

In our struggle with the contrary powers it is not only with the cleansing of the vital-energetic fields of our own Holon that we should be concerned. The purity of the emotional (astral) plane is far more in question. This is where the vulnerable realms are located through which alien powers can infiltrate.

In critical situations the following exercise has helped me come back to myself and return to emotional clarity:

- Find your mid-point—it is deep in the chalice of your hips, roughly where the lumbar channel is located. Rest there in your mid-point for a while, wholly strong, decided and present. It is worth recommending that you ask for support from your spiritual master or guardian angel the Archangel Michael or another spiritual helper.
- When the right moment arrives, use your imagination to let a sound like the Big Bang spring out from your mid-point. It is important that this should not be a fiery sound but "cold"— as if a piece of iceberg had broken off and fallen into the polar sea. It is important too that you direct the sound of your Big Bang inwards and not outwards.
- Following the Big Bang, cleansing waves spread like rings outwards from your mid-point. To heighten their purifying power, you can imagine these concentric waves to be bathed in violet light. When they reach the rim of the Holon and are flung back, their color is changed to white in order that they may send the quality of purity through the Holon.
- If necessary, repeat this exercise several more times. Give thanks for the spiritual support that is given you.

CHAPTER SIX

# The Consciousness of Earth Organizes Anew Its Plane of Feeling

### Elemental Beings in the Service of Charitable Love

YET ANOTHER LAYER OF EARTH changes was set in motion during the second half of 2002. It paralleled the unpleasant strengthening of the "dark powers" but was more auspicious, radiating joy and optimism.

On June 23, 2002, a dream of such a dramatic nature that I was struck with fear pointed me towards this new layer. However, it turned out that the threat was only for show, so that the dream's message should receive my undivided attention.

In the dream, I am invited to participate in a publicly advertised call for tender. After handing in my project, I am convinced that I will be awarded one of the bids. I am all the more unpleasantly surprised by the curt note from the jury that my plans have been drawn with too faint a pencil and could not be photocopied. This is the reason why my project is rejected. I can come by at any time and collect the papers. When I enter the office, I am shown two sheets of drawings. They are full of rhythmic patterns, painted in very subtle, pastel-like colors. Next I am handed a plastic bag. Inside I see two sheets of paper, which, because of the failed attempt at photocopying, are wholly black on one side. Now it's going dark before my eyes too, because I feel I am being done an injustice. Should a project be rejected just because the drawings are so fine that they cannot get through the coarse process of photocopying?

In the light of later revelations it can be seen that the dream is announcing a new phase in the process of Earth change. It is emphasizing a particular fact that has obviously taken on decisive meaning: that I am not sufficiently finely attuned to perceive the qualities that have become the focus of this new phase of evolution. The subtle patterns on the two sheets of paper are an indication of how tender and gentle are the powers involved in this new phase of Earth change.

First, however, I had the following dream that initiated me into the mystery surrounding the original source of these highly sensitive powers. I am in the midst of packing my bags for a lengthy journey. The suitcase is nearly full when I see a thin, dark-skinned hand slide out from among my pile of shirts. At first I think that an animal must have hidden in my laundry. But then, that little hand might belong to a small child—though there's not enough room under my shirts for a child's body. In addition, there is a hint of webbed skin around the fingers, so this cannot be a human child.

The thin little hand lays itself lovingly on my arm as if wanting to banish my fears. Wishing to discover who is playing hide-and-seek with me, I press lightly on the stack of shirts beneath which the being, whatever it is, must lie hidden. I touch something soft and silky like an infant's belly.

I awoke from this dream feeling very blessed, and was at once reminded of the tender feelings that had been embodied in the two drawings from the previous dream. They had radiated an indescribably tender and, at the same time, well-grounded love. It was this very same quality that I had felt dwelling within the little hand of the unknown being. I had had a similar experience a few times previously when I had been in contact with one of the "new," changed elemental beings.

The most impressive experience of this sort had happened a year

## The Consciousness of Earth Organizes Anew Its Plane of Feeling / 159

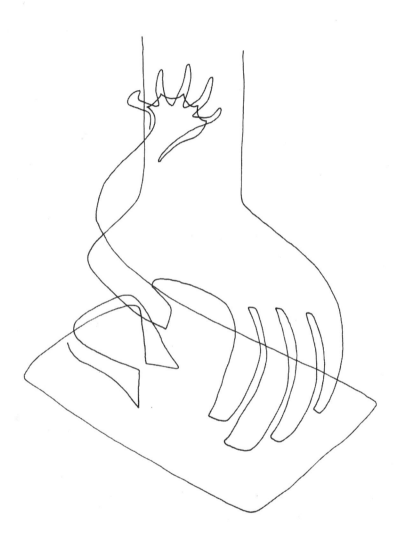

*A thin, dark-skinned hand slides out from the pile of my shirts and touches my arm.*

earlier in the Austrian Waldviertel when I was giving the seminar at Rastenberg, as previously mentioned. I wanted to "show" the seminar group an elemental being that is attuned to Earth's new form and vibrates in harmony with it. Close to the seminar house I found two of these elemental beings.

I perceived one of them in a lonely clearing and at once felt interested in what its task might be. In answer, it showed its face, and I saw a snake rising erect at the site of its third eye. The image was accompanied by the quiet explanation that it was a messenger whose task was to spread the news of the resurrection to the elemental world.

I might have had no inkling what resurrection meant in this connection if I had not encountered Pan earlier—an event that occurred in a forest in Saarland during the spring of 2000 and which I have already described several times. There appeared before me then the tall figure of Pan, the ancient God of Nature. I could plainly see the marks of the wounds of Christ on his hands and feet and also in his side. Silvery rays shot from these marks to the corresponding points on my own body.

In my book *Daughter of Gaia* I set forth Pan's message as follows: "The Son of the Virgin has died to the human world to reappear through nature."

It follows that we are here looking at the renewed appearance of Christ in the context of the present Earth changes. We have already spoken of his appearance in another form—as a complementary manifestation of the same cosmic power—in connection with humanity's cosmic double. In that case the return of the Christ is accomplished through the inner work of many individuals who are attuning to the core of their own being and transforming the shadows of the past.

The way in which the Christ Power is now revealing itself though the beings of nature is quite surprising. In this mysterious process the elemental beings undergo an inner change that can be compared to the resurrection of the Christ. Previous to this, they were entirely bound to their own world and specialized field of activity. After their resurrection, they have become the bearers of cosmic love, which is the same power as has wrought the change in them.

The second of the new type of elementals that I met at Rastenberg will help us define more precisely where we are being led by the change in them. This being was located in another spot close to the seminar house. Scarcely had I turned myself within to connect with it when my heart center was changed to a fine-cut ruby. Mere contact with the elemental being had enormously increased my capacity to love.

Moreover, I felt a brotherly closeness to the elemental, as if it were my companion being. The experience of this close relationship between humans and elemental beings can be traced to knowledge of the Christ, which is now entrusted to us both. This is the same divine revelation that in the West we relate to Christ's incarnation. It is a revelation of all-embracing love and also of individual creative power. Obviously, it is no longer confined to the human world only, but is also extended to other beings of the Whole.

To these qualities can be added another that I have discovered among the "new" elemental beings. It is comparable to human individuality. It appears as if, during the course of their renewal, the elementals become individualized in a particular way. Beforehand, they were completely engrossed in their tasks within the framework of the life processes of nature. There was no personal free space available to them, such as we humans know.

Admittedly, they do not yet know the personal "I" in a human sense. Every elemental being is still like a little stone in the all-embracing mosaic of Earth. However, passage through the Christ initiation has enabled them to develop a collective self-awareness. This makes it possible for them to take the next step and reach the level on which they can communicate with humans. Until now, I have only experienced a similar closeness to the human world among those rare elementals that have worked in the past on performing particular rituals with humans and were thereby initiated into the world of human qualities.

When this second elemental at Rastenberg was asked about the task assigned to it, its answer was also characteristic. It let me know that its path of service lay in imparting a blessed feeling for the closeness of nature to anybody who might pass near. It added that it was itself a transformed elemental being of the fire element.

This is the precise point that is important for our understanding of the new elemental beings. These are not really new elemental beings, but old elementals that have gone through a deeply wrenching process of change, which I compare to a spiritual initiation. Some elementals do not exhibit this new quality, either because there is no need for them to do so or because people are still not open to it.

In this connection, my contact with the nymph of St. Gallen, Switzerland, during September 2002, may be of interest. The nymph dwells at the spot where a wild stream reaches the townscape of St. Gallen, whereupon, unfortunately, it immediately disappears into canalized conduits. According to legend, it was from this spot long ago that the town started to develop. The canalization is also the reason I have known the nymph for a long time; it is some years now since I began leading regeneration seminars through the townscape of St. Gallen.

When I was paying another brief visit to St. Gallen in September 2002, people told me that they thought the nymph was in difficulties and asked for my opinion. I went to her historic dwelling place to get a picture of what was going on. My first perceptions of the nymph were in fact unpleasant. In some way she appeared broken. The whole area of her "head" was covered in dark shadows.

I had already started to worry about her when it occurred to me to change my point of view. Should I not attune myself to the plane of her "new" elemental being? From there, I found I could perceive a quite new and welcome aspect of the nymph. She resembled the face of a loving woman, which was so large that it could pour itself out from that spot through the whole town of St. Gallen. The presence of the nymph was no longer in the exclusive service of the realm of nature, but also conveyed a blessing for the people who had set-

## The Consciousness of Earth Organizes Anew Its Plane of Feeling / 163

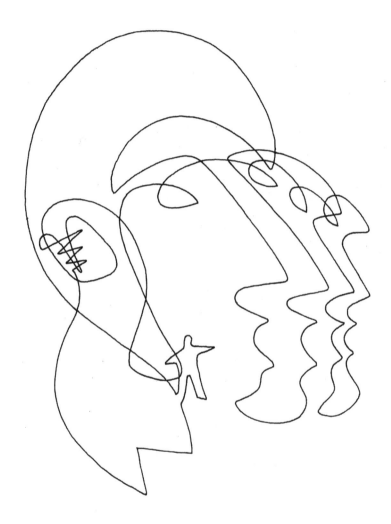

*On the left, the darkened figure of the old nymph of St. Gallen; on the right, her new, expansive presence.*

tled there. On the one hand, her tragic garment mirrors the wretchedness of the alienated civilization that has spread around her feet. On the other, she tries to inspire people to go through the change that she herself has experienced.

You will certainly be interested in the exercise that helped me attune to the new aspect of the nymph of St. Gallen. It was given me two months before my meeting with her, and it happened in Mondsee, Austria, when I came by chance upon a group of the changed elemental beings.

I was staying in Mondsee for a seminar in nearby Salzburg, and it was evening when I arrived there. Despite the lateness of the hour, I wanted to pay a brief visit to the lake of the same name, for I had been actively using Earth healing through the whole year to promote its well being. As I walked toward the lake along a line of giant lime trees and was already near its shore, I suddenly stopped and stood still. I had come upon a group of invisible beings.

To perceive elemental beings, I often use an exercise that starts from the chakra of the Earth element, located between the knees. If we humans had a beautiful tail, like a she-wolf's, for example, this chakra would be located at the end of the tail. For perceptive purposes, one concentrates on the chakra between the knees and at the same time opens oneself to communication with the relevant being.

When I tried to use this method at Mondsee, my focus, fixed on the chakra between my knees, was guided upwards. The perceptive point was now located between the belly and the heart. When I opened myself from that spot to the presence of the unknown beings, I could perceive them with unexpected clarity.

This was a group of transformed elemental beings that had placed themselves in the middle of the path along which countless tourists walk daily to reach the shores of the romantic lake. The people who take this path go unhesitatingly through the midst of the group. In doing so, they are lovingly touched by the "new" elementals and are inwardly calmed and inspired. The invisible ones demonstrated in my own person how capable they are at filtering out the energetic and emotional garbage from the auras of passers-by and transforming such trash in the light.

This example illustrates the progress that elemental beings have made over the last three years in their change process. A dream that I had during the night of July 21–22, 2002, which preceded my encounter with the "new" elementals at Mondsee, foretold the next phase of their transformation. Its message came to full realization during that same summer of 2002. This was how the dream went. A small group of us are standing beside a much traveled thoroughfare. We want to cross the street, but the fast moving, tightly massed traffic does not allow it. My daughter Ajra, who loves riding, is also there, and she is holding the reins of a large and handsome horse. Now the horse will wait no longer. It rears onto its hind legs and leaps forward, its forehooves landing on the car that is just then passing us. The car is an elegant, silver-colored limousine driven by an elderly gentleman. The horse performs the dangerous jump so dexterously that there is no real damage caused, though unfortunately its forehooves have brushed against one side of the car. There is stable manure clinging to its hooves, and two thick, stinking stripes stay stuck to the car's side.

In some consternation we watch for the driver's reaction. His limousine has stopped a little distance away on the other side of the street, and he is staying in it. Two women from our group are running towards him, each holding a bottle of wine in their outstretched hands. I wonder whether they hope perhaps to calm the driver's anger with the bottles of wine.

In the dream, the cars roar past us, their drivers sitting on their own at the wheel. They represent modern human beings who are isolated and confined in their mental armor and constantly pass reality by. "My" group at the side of the road is made up of human and elemental beings—all conceived in the above mentioned sense of our new relationship with the elemental realm. Not only the horse but also the rider holding its reins point to the world of elemental beings. It is appropriate for my daughter Ajra to take on this role.

Her book *Von der Ewigkeit Berührt* (Touched by Eternity) has vouchsafed us entirely new insights into the elemental realm.

The dream tells us that it is not the human beings but the elementals that have taken the initiative in this new phase of Earth change. The powerful horse that has manifested two thick streaks of excrement on the limousine's side is one illustration of this. Another is symbolized by the two bottles of wine that are intended to soothe the driver's anger.

I take the two streaks of manure left on the limousine's glittering surface to be a warning. I relate this to the natural catastrophes that are occurring ever more frequently in our environment. The month following the dream was particularly significant in this connection. On August 14, the Moldau River flooded Prague; 45,000 people had to evacuate the city. The great flood reached Dresden a few days later and afterwards other cities down the River Elbe. At the same time huge forest fires were raging on the other side of the world on the island of Borneo.

At that time I was on the little island of Srakane in the Adriatic Sea, where I retire for a while each year to meditate and study. This gave me the opportunity to consult once again with Julius, my master from the world of elementals, who has dwelt there on a little hill for time out of mind. The question I put to him was this: "Natural catastrophes are something completely natural. They have always been natural and will be so in future as long as the principle of change is part of the evolution of life. How is it that of late the news of dramatic upheavals in nature have affected many people so deeply? It is as if a message were somehow interwoven with the events."

The master pointed to the signs that horse's hooves had produced on the limousine. Yet the vehicle was in no way damaged! So the signs cannot be seen as nature's assault on humanity's egocentric attitude. Rather, they are a message. But because the message is itself alarming, it is always conveyed by the sort of dramatic event that nature can evoke through its four elements. It can dispose of earth-

*The horse, representative of the world of elemental beings, will no longer wait at the roadside.*

quakes (the earth element), volcanic eruptions (the fire element), hurricanes and whirlwinds (the air element) or raging tides and floods (the water element).

Elementals stand outside the human beings' indoctrinated world, and they can very well perceive the disastrous state in which we moderns find ourselves. Encapsulated in the "limousine" of our perfectly crafted thinking, we are on the point of losing our last contacts with the reality of nature, earth and cosmos. And so we steer blindly into the abyss.

Nature does not have the capacity to debate the threatening danger with humanity. The language of direct physical experience, conveyed through natural catastrophes, remains the main recourse open to elemental beings trying to warn us against the abyss into which we are about to plunge. So it is by sensitively disturbing the usual circumstances of life and allowing the resulting grief to happen that their alarming message may possibly penetrate humanity's mental armor and reach our consciousness.

However, the elemental beings have elected to help humanity in other ways than by sounding the alarm at regular intervals. Another possible way is indicated in my dream's closing scene, where two "women" from our group hurry to the angry car driver, each holding a bottle of wine in her outstretched hands.

I had wondered why these two women appeared so completely identical. This almost never happens with human beings, but among elementals the secrets of individuation are not so well known. So it must be that these were also elemental beings and concerned to help a person in distress, a possibility hitherto unknown to me. In that case, why were bottles of wine the symbol of help? Could the man be intoxicated?

In answer, old, wise Julius drew my intuitive attention to a small, stunted pine tree. I stood there quietly for a while. Suddenly the indwelling elemental, which I had known for a long time, began to "milk" my sex. A blissful feeling spread throughout my body. What could this mean?

Julius directed me to the human body's mood-lightening neurotransmitters known as endorphins. They are, for example, released in the sex act. But what have they to do with elemental beings?

The answer told me that this matter does not concern the four elements, powerful beings that guide natural events and sound the alarm by initiating natural catastrophes. I learned that instead it concerns the mission of the personal elemental being that accompanies every person through his or her incarnation. This was already mentioned in the first chapter.

The personal elemental being finds itself very much at ease in the realm of the sex organ and can release the stimulating neurotransmitters of its own accord. Here we find the meaning of the symbolism of the wine bottle. Endorphins can be spilled and so help people mitigate rigid, firmly held thought-patterns. Afterwards, those persons will be better able to follow the promptings of their heart and live more lovingly.

The elemental beings that offered the man "wine" to soothe him symbolize the possibility that the powers of nature will appeal to individuals from within. Julius assured me that there would be no attempt in such a case to break through the mental armor in which we hold ourselves prisoner. Instead, the personal elemental being is much more committed to "soften" and "release" the latter from within. To show me how this method operates, the following exercise was proposed:

- Center yourself in the midpoint of your heart. Then direct your attention downwards along your spine till you reach the endpoint of the coccyx. Next, galvanize the power concentrated at the base of the spine by rhythmically moving your two hands behind your back. The movement is like that of a dog wagging its tail (A).
- Intersperse this movement by frequently raising both your hands to grip your head and then, as if they were a comb, drawing them down the length of your spine and then up

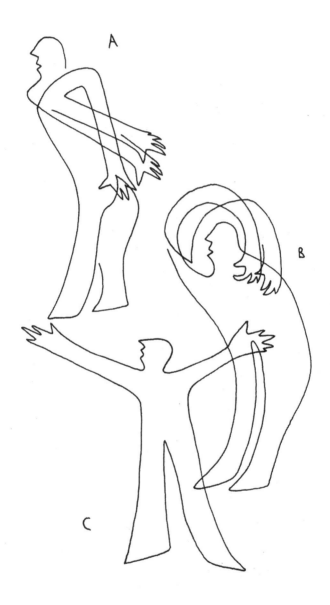

*The artful exercise of wagging one's tail.*

again (B). This movement spreads the impulses released by the elemental being through the entire body (C).

However, it does not suit the innate self-sufficiency and individuality of the human being to be dependent on elemental beings to ground our loving feelings. The more I was inspired by the activity of the personal elemental, the stronger was my desire to find an exercise through which every human being would be able to awaken his or her power of elemental love. In answer, I was given the following simple exercise:

- Imagine that a tiny golden fragment breaks away from the mandala of your heart center and glides slowly downwards through the watery layers, i.e., the emotional layers, of your belly. Finally it comes to rest on the floor of your pelvis at your pubic bone.
- Prepare yourself to get a precise sense of what happens when that little golden fragment touches the pubic bone. Follow the feelings that will quite possibly explode as they strive to rise upwards. Let your heart bathe in them so as to draw into itself the power of grounded love in all its wholeness.

The next question is what is meant by "grounded" or "elemental" love.

The answer involves removing the barrier that divides human love energy into two separated parts. The one part is equated with the sexual drives, the other with the spiritual dimensions of love. This disastrous separation weakens people's ability to move mountains by using the power of their love.

Much has been spoken of love's power, but the causal world turns ever onwards, lovelessly and destructively. To overcome the heart's powerlessness, we must unite the well-grounded elemental power of love "below" with the spiritually exalted power of love "above." This brings into existence within the human being the unified emotional field of heart power—and from this comes a real hope that the world can finally be reshaped around the laws of love.

The fairy of a little olive grove on the island of Srakane showed me a second exercise by which to ground the power of love:

- Hold both hands below the heart center, horizontal and back to back.
- Imagine that your heart center, in the form of a golden ball, is lying on the surface of the upper hand.
- Slowly lower both hands, still carrying the ball, till they reach the region of your sex. Let the golden ball stay there for a while and unite yourself with the potential of the elemental power of love.
- Hold the lower hand near your sex while you raise the upper hand till the heart center is back in its place.
- Try to sense the power field of grounded love that has built up between the lower and the upper hands. Take it into your whole being. Breathe with it.
- To further strengthen the heart field through polarization, you can lay the upper hand alternately on the front and back sides of your heart center—the back is located at your back—and perceive the change in relation to your lower hand.

**Special Changes in the Consciousness of Earth**
I had another question to ask of old, wise Julius. "Have elemental beings changed only in regard to their relationship with human beings or also in regard to the other natural phenomena?" There followed a brief silence. Out of the silence I received a clear instruction to stand up and walk in a certain direction out into the countryside. I let the master's impulse guide me, and at a certain point it disappeared. I had arrived at the desired place.

I had known the spot ever since, years earlier, I had conducted a geomantic investigation of the island. One of the acupuncture meridians of Earth, i.e., a ley line, runs through this place and joins

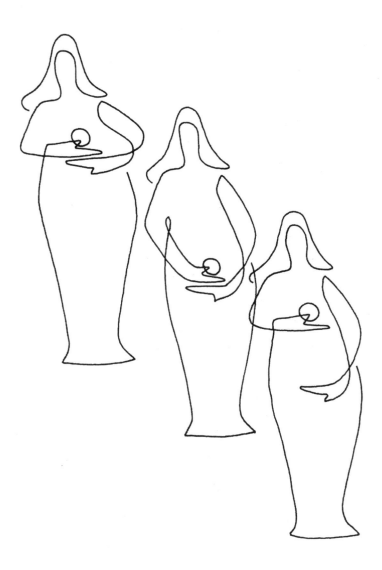

*The holographic exercise for grounding the power of the heart.*

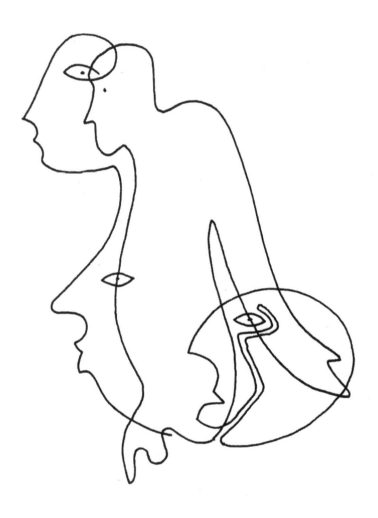

*Opening the three eyes of the soul—an exercise.*

Istria in the west with the Adriatic coast in the east. I had obviously been led here to view the energy line afresh. Therefore, I opened the eyes of my soul in order to perceive the course of the ley line, invisible to the outer eye.

Have I already mentioned how one opens the eyes of the soul? The phrase "eyes of the soul" puts one in mind of the third eye located behind the forehead. It is certainly correct that this is the soul's organ of sight. However, the soul is not a being with human properties, such as strives to concentrate its vision on a single point. This is the reason why the third eye, which is really a spiritual organ of perception, is imagined to be a leftover from the old patriarchal epoch of human evolution.

The soul represents a fractal—a holographic fragment—of the Goddess. Consequently it displays three aspects, which are:

1. The holistic aspect of the soul (the maiden)—color white.
2. The creative aspect of the soul (the partner)—traditional color red.
3. The transformation aspect (the old wise woman)—color black.

One can further imagine that each of the soul's three faces is provided with an eye. It follows that the soul has not one but three eyes. The three eyes are each assigned to a spot in the physical body that is characteristic of one of the soul's aspects:

1. Behind the middle of the forehead: the eye of the Virgin Goddess within us.
2. At the upper edge of the belly region: the eye of the Partner Goddess within us.
3. At the back, underneath the sacrum: the eye of the old wise woman within us (like the others, it faces forwards).

The complete exercise is to be found in the Appendix (see page 241).

When I looked at the energy highway with all of the soul's three eyes, I discovered a thin, tall being that was running along the ley line. I had a sudden impulse to bless this being, and my hands made the appropriate gesture. At that moment the being perceived me. It threw itself head downwards into the stream of the ley line and stretched its "feet" upwards to the sky so that it looked like the letter "Y." The sign was multiplied countless times and distributed along the energy highway. I gathered this meant that from now on there is a consciousness dwelling within a ley line, that raises it out of the status of a pure energy phenomenon. As a generalization, this means that all energy phenomena have received a strong consciousness component through the course of the present Earth changes. They are incorporated with the "new" elemental beings.

On the very next day I found another instance of the change in the elemental beings of the countryside. This concerned the old giant that had his place behind Skrakane's abandoned school building. I had met him for the first time ten years before and reported on the encounter in my book *Nature Spirits and Elemental Beings*. At that time the giant was staring uninterruptedly into the distance, and I had experienced him as a relic of a long past epoch of Earth's evolution. It had not been possible to find out at the time whether he was also fulfilling any present function.

This time I was surprised to perceive him full of life. Instead of the rigid, tower-like body, I now sensed a dynamic double loop that pulsated between heaven and Earth. Furthermore, the old giant no longer appeared as if closed up within his own energy structure. His essence shone far out into the surrounding area. The "old ruin" had been reinvigorated by a new task.

To learn more about the old giant's new task, I turned telepathically to the Master Julius. He replied that for long whiles past the old giant had represented a part of Earth's consciousness that was slumbering and held in reserve. This was the state in which I had perceived him years before. Because Earth changes are a complicated process, Earth must activate many new functions of her con-

## The Consciousness of Earth Organizes Anew Its Plane of Feeling / 177

*The old giant on the island of Srakane had a new life as a kind of nature angel.*

sciousness in the course of them. Her reserves are also involved in this, and the old giant represented one of these reinvigorated reserves. His function had been changed to that of an instructor to the elemental world. Julius conveyed to me that his task now consisted in teaching the beings of the individual elements about the new nature and consistency of space. He added that in this respect one should bear in mind that, for elemental beings, to teach was to be.

Later there followed a dream through which I learned about a special tool that the Earth consciousness has developed in order to be able to intervene in the human world. I call it the "threads of peace."

First, I am shown an elderly man who resembles a shaman and therefore suggests some special closeness to the elemental world. This also tells me that the dream revolves around the theme of elemental beings. Next the old man becomes an ordinary person who falls into a fit of rage. He shrieks and throws his arms and legs about. The situation is so serious that the police are called. A policeman enters but does not have his pistol with him; neither does he have his mobile phone to call for help. The situation appears so extremely threatening that I see no way out but to hide from the truth and pretend it isn't happening.

Translated into logical language, this means that in future we should reckon on violent struggles occurring within the ranks of humanity, and when this happens the traditional methods of safeguarding public order will no longer serve.

The dream next shows me a picture of the tool that the elemental world has developed to help the people they love get through the anticipated turbulence. The man who had the fit of rage is now lying on a wooden stretcher and is carried past me up a staircase. I am surprised to see that he has become perfectly calm and seems as if sleeping. As I look more closely, I realize that he is bound tightly to the stretcher by the thinnest of thin threads. Obviously it is this that has succeeded in calming him down. The thread is transparent and as thin as a spider's; it could never bind that furious man to the stretcher in any physical way.

I make the association with a magic thread. Such a thread would have a special role in connection with any sort of love magic. If a couple is bound together with an invisible thread like that, it is extremely difficult to separate them.

However, that association fits only part of the dream's core meaning, since here we are dealing with a magic thread which establishes peace. To learn how the "thread of peace" is woven by the elemental beings, I had to turn to Julius once again. He sent me down the hill to that entryway to the underworld with which I had already been made familiar when I was writing my book about elemental beings.

The Earth elementals whom I met there were dwarfish in appearance. The information they gave me related to the way in which the thread is used. If, for example, a person is facing some very aggressive threat, he or she can ease the situation by spinning an invisible thread and putting it around the happening, whatever it is. I asked the dwarves whether humans too could use this wonderful technique. Yes, I heard them say, it may be possible, provided that sooner or later the relevant crisis is tackled and transformed. The threads of peace hold only for as long as they are absolutely necessary.

Afterwards, I asked the dwarves whether a human being could spin the thread, and if so, whether I could pass on the knowledge. I was told it was permitted, provided it was not used for mischievous purposes.

With this assurance my two hands were initiated into spinning the invisible thread:

- Hold your hands in front of your chest. On each hand the thumb and ring finger are touching each other.
- Next, move the two hands rhythmically up and down. The upper hand goes down and the lower hand goes up.
- As the hands glide past each other, relax the thumb and ring finger so that in that moment they do not touch.

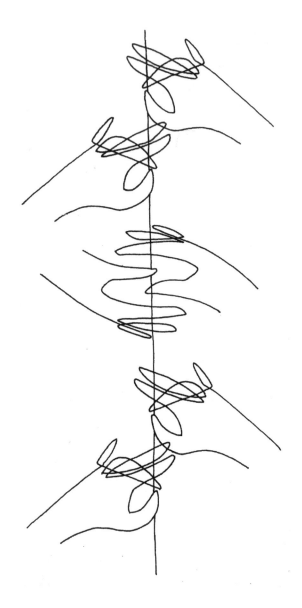

*The holographic exercise of spinning an invisible thread, by which you can bring about peace.*

I should like to mention another meeting with the "new," changed elemental beings. At the end of October 2002 I visited the city of Dresden, which two months before had suffered severe flooding. In the evening I was to give a talk under the heading, "What is Happening with Earth and Her Elementals?" and for the purpose had undertaken to make a tour of the city with a group of colleagues to get a picture of the consequences of that natural disaster.

We were all agreed that there was a wonderful atmosphere pervading the city following the flood. The River Elbe no longer felt herself confined to her previously prescribed river bed and was pouring the unique quality of her feelings freely into the surrounding cityscape. It felt as if her presence was still as broad as during the catastrophic flood of the previous August. In a physical sense, the Elbe had certainly returned to her original bed, but not in the sense of her emotional and spiritual presence within the city and landscape.

Finally I left the group and went alone to cross the wide meadowland that separates the city from the river. I wanted to get in contact with the nymph of the River Elbe. There is a particular exercise I prefer to use when I wish to make a connection with the elemental beings of the waters by coming into resonance with them. I visualized and actually sensed that there were fish scales on the sides of my thighs. In addition, I imagined that my feet had grown together to form a fish's tail.

When I had finished making all these preparations, a beautiful young woman swam into my vision from the depths of the river. Telepathically, she gripped my fish's tail and drew it forward. Then she let it glide slowly upwards, tight against my body. The fish's tail slid upwards till its fins came as high as my heart chakra. It felt then as if my heart's center was being held in the chalice of the fish's tail. I realized that I was being shown the cosmogram for communication with the changed beings of the water element.

Then another woman of the waters, much more powerful, emerged from the river deeps and spoke to me. I can remember only two of the three statements she made. Translated into human words,

*The Nymph of the Elbe River has guided my fish's tail up to my heart center.*

the first ran as follows: "We beings of the elements have created the wonderful natural world that surrounds you. Afterwards, you developed many splendid cultures. Now we have become equals. Let us be partners in creating a new civilization!"

The second statement sounded like a warning directed at humanity: "In the last analysis we represent two strands of one and the same Earth evolution. Why do you shy away from us; why do you keep yourselves encapsulated in your own pods?"

### The Transformed Ego of Humankind

With that last statement the nymph of Dresden had put an important key in my hand. If it is the case, as was indicated, that our earthly incarnation really couples us humans so closely with the evolution of elemental beings, then is it not logical to assume—whether or not we personally want it—that we are being carried along with them through the self same changes that they have experienced over the past few years.

If this is the case, we should envision how we may regard our relationship with the evolution of elemental beings in two separate ways. There is an objective as well as a subjective point of view. Both have fundamental significance for the human being involved in the change process.

The personal elemental being that accompanies each one of us through our entire earthly life, as portrayed above, belongs to the objective aspect of our relationship with elementals. In the mother's womb, the personal elemental being is already helping to build the body of the child. After the death of "its" human, the elemental being remains to serve at the side of the dead person until he or she has succeeded in laying aside the last of the fleshly veils to go free into the kingdom of the soul.

Our relationship with the evolution of elemental beings is undisputable and can also be seen from a subjective point of view. However, in this case, one cannot talk of the personal elemental as if it were a being separate from the human. Since, in the course of its history, humanity has become a component part of earthly evolution, each one of us is also unavoidably integrated into the great community of elemental beings. Or, to put it another way, we have become a component part of the consciousness of Earth.

How does this "unavoidable" relationship manifest—aside from many people resembling dwarves or fairies?

The best answer comes by getting up close and personal. I suggest an exercise that I have found reliable and have tested many times with a variety of groups. My observations suggest that if whoever is doing the exercise does not feel or sense something, it is because he or she is self-conscious about his or her own elemental aspects, which we humans have ignored through long epochs of our evolution.

This exercise can best be performed while sitting in a chair, since one then has four additional "legs" at one's disposal. However, it can also be done while lying on a bed. Because we are so perfectly at *one* with the elemental being throughout our entire life, it is scarcely possible for us to perceive it. The exercise is so arranged that first, with the help of the imagination, it puts one at a distance from one's elemental being, and immediately afterwards creates a reunion. The moment of reunion must not be delayed, because only then is the inner elemental being consciously perceptible:

- Imagine that you are sitting on a mirror. When you look down, you see yourself—your second self. If you are doing the exercise while lying down, imagine that the mirror is leaning against the soles of your feet.
- When you are ready, let the second self—the one you see in the mirror—stand up quickly; you then draw it up to your shoulders.
- Embrace it so that your left hand lies on your heart center and your right hand on your solar plexus.
- How does the presence of the elemental being feel to you? What quality does it bring to the Holon of your being?

## The Consciousness of Earth Organizes Anew Its Plane of Feeling / 185

*An exercise for meeting with one's personal elemental being.*

Caress it for a while; express your thanks or make your request.

In the first chapter I described how I was given a gift for my birthday in 2001: the revelation of the being of Luz. I was shown one of the changed elemental beings that serve as midwives to the whole being of humanity. On my next birthday too I received a gift from the elemental world: a dream gave me an insight into the newly forming relationship between the personality and the elementary essence in human beings.

I dreamt that early one morning the mailman surprises me by delivering a book. As I open the package, I see that the cover bears my name. It is therefore to be expected that the book will contain my work. But there's no trace of it! Works of art with explanatory commentary are pictured on every page, but I can find among them no single example of my own work. My ego gets edgy and feels deeply hurt. To satisfy its ambition, I look for a picture that one might "adopt," based on a similarity with my work. Then I notice that two pages are bound to the back of the book by a fragile web of skin. The webbed skin is the same as I had perceived on the elemental being's little, loving hand when it reached out from amongst my baggage.

The dream caricatures my personal reactions, an indication that this has to do with some change in the sphere of the ego—something to which Western people always react with extreme sensitivity.

The dream is obviously dealing with the future loss of the egocentric personality, which is usually also described as the ego. Moderns do not want to hear about the loss of ego because the ego is seen as the protector of personal identity. Is a human being still human if the individuality is surrendered?

However, we are not thinking about the cancellation of human individuality. That is a solid component of the human being. It is part of our humanness that each one of us should be centered in our unique and individual wholeness and should fulfill our personal role within the universal Wholeness.

Of much more concern is that in the course of their long alienation, the "I" of the modern human being has split into two. On the one hand, the "I" of the spiritual soul has been enclosed in the pod of alienated personality. This has almost completely silenced the voice of the soul and the "higher I." On the other hand, the external "I" has been gradually developing itself, and during the supposed absence of the inner voice has taken over leadership of the whole.

This coup has caused problems. The externally oriented egocentric personality knows nothing of the archetypal images that would lend some deeper meaning to its actions. Because the ego routinely ignores the voice of the soul, the mind gets substituted in its place. And at this point we hit up against the second problem with modern egocentricity.

Because the mind works incessantly on the plane of logic, it is incapable of answering questions regarding spiritual guidance. It follows that the "I" is forced to listen to the invisible signals of its own surface emotions. There appears no other way out. In consequence we humans are continually being drawn into chaotic situations.

The dream about the book with webbed skin on the back cover pointed to the unlooked-for possibility that the personality can be changed without losing the degree of individual self-sufficiency that it has achieved. The dream presents the water element as the future foundation on which the renewed personality of humankind can unfold. The water element speaks to the emotional world. It is not in the airy world of thought that our ego will find its new anchorage, but in the watery world of feelings.

It is in this sense that we can understand the statement made by the nymph of the River Moldau, whom I had questioned in September 2002 after the great flood in Prague. She had closed our conversation by saying that, apart from anything else, it was time for people to learn to use their gills.

One could think this a jest, but the words were spoken with great earnestness. This fact later moved me to ask the elemental spirit of the River Rhine about it. It was during a railway journey along the

Rhine that I tried to make a connection with the being of the river and he appeared to me. The contact was made between Mainz and Coblenz, where the line runs close to the riverbank.

I asked the elemental spirit of the Rhine where one should look for gills in the human being. In answer he began to plait a wreath, formed of several strands of water, around my heart center. The plaiting started from my vocal cords and moved from there towards the left side of my chest, where it passed through the left-hand chakra of the water element and continued towards the middle of my belly. After transiting the right-hand chakra of the water element, it then moved back towards my throat, and the "water wreath" was closed off.

The input from the spirit of the Rhine can be transferred to an exercise. With its help one can track the site of the personality's future home. Like Snow White in the famous fairy tale by the Brothers Grimm, the elemental beings are offering the personality a place of refuge. However, the refuge is not in the dwarfish realm of Earth but in the watery wreath of the personal elemental being within the human:

- Find your inner silence.
- Make contact with your elemental essence while raising your attention slowly and lovingly from the sexual region towards your midmost heart. Somewhere between the solar plexus and the heart's center you will find the point where your elemental aspect has made its home since it joined in the current changes within the elemental world. Rest there in its center for a while.
- Starting from the area of your larynx, you now begin to braid the wreath of water. It is braided out of several strands of crystal clear water. It runs across the left-hand chakra of the water element down to the middle of your belly, and then across the water element's right-hand chakra back to your larynx.
- After that, direct your attention to the above mentioned center, which is now also the center of the wreath. You are pres-

## The Consciousness of Earth Organizes Anew Its Plane of Feeling / 189

*The gills of the human being: a water wreath that reaches from the larynx across the two chakras of the water element to the solar plexus. Midpoint is the heart.*

ent in the middle of it and you let the wreath around you become round too. You allow its watery substance to transfer itself into the feeling of grounded love that the elemental beings know.
- A watery, spherical space arises in the middle of your Holon. In it are joined the forces of head and mind, as well as the qualities of belly and feeling. Henceforth, whatever you do in everyday life, you will be learning to think, will and act out of this new space of integrated personality.

On September 16, 2002, I was given another dream that made the ways of changing the ego even clearer. In this dream I am standing in a flat meadow, waiting for the other members of my family. I am constantly looking into the distance behind me, as if I am expecting members of my family out of the past to be coming along soon.

In this I see the image of the old egocentric personality that lacks the capability to see into the future but is dependent on what the mind has stored in the shape of memories from the past.

Suddenly the meadow behind me stands upright. It makes a wall, destroying any hope that other members of my family may be following me. I feel absolutely and finally lost, isolated and cut off from my accustomed roots.

In this I am experiencing the personality's sense of helplessness in the face of repeated somersaults of etheric space that allow the emergence of completely new living environments.

Quite unexpectedly, a little door opens in the sky. There is a wooden ladder leading to it. My two daughters leap out of the open door, weeping so copiously that their tears stream down into the depths. They tell me that their mother is sick. The news, supposed to be sad, in no way affects me adversely. A wave of happiness begins to stream through my being. I experience myself raised up once again in my family circle.

The news of my wife's sickness appears rather to be a symbol of change. Nor do the streaming tears have anything to do with sorrow.

Instead they represent the new function of the emotional plane, as related above. And finally, there is a wooden ladder leading to heaven, which I can climb to reach my loved ones once again.

Taking all the images together, the dream sent me a message that the days of the old personality model are over. In any case, it is only the poor foundation that supports the modern ego that will vanish through the repeated somersaults of the space we inhabit. However, I do find comfort in having been granted an insight into the new configuration of the human personality.

If one looks at both dreams and the messages they bring, one can characterize the shape of the new born human personality as follows:

- The foundation for personality's new shape rests on the changed elementary essence of individual men and women, an essence that is filled and overflows with charitable love— and has the goal of ennobling the ego.
- It is the path of constant change through which the personality evolves towards the holistic human being—in whom the different levels and dimensions are no longer separate.
- From now on, the watery dance of the emotions forms the foundation on which a person's individuality and personality shall be expressed.

CHAPTER SEVEN

# The Next Goal of the Processes of Change

## The Arrival of the Redemptress

READING A BOOK CAN BE seen as a form of communication between yourself as reader and me as author. Viewed in this way, a book can be interpreted as a psychic space for the exchange of ideas, experiences and knowledge.

Taking the thought further, one can also imagine a book to be a space of initiation where author and reader are initiated into mysteries concerning which neither previously had the slightest premonition. I am sure that this is generally why one begins to write and read certain books. I am interested in something more: Could a book also be configured as a creative space where author and reader can work together to change Earth and humanity? Just the act of reading a book brings about an inspiring form of co-working anyway. My words set off certain triggers, one after another, that move my partners—my readers—to think and feel along the same lines, and also to develop contrary interpretations. This sort of activity is also called the creative process.

Consider that this book will be read by several thousand men and women; each time the co-creative process will be set in motion afresh and may even be varied. Practical exercises are embedded in the text to ensure that the process of change remains grounded in personal experience. The drawings should prevent the impulse from wasting away in arid avenues of thought.

This idea of mine, that by reading a book one can share in this epoch-making process of change in Earth and humanity, is based on an impressive vision that was granted me in the year 2000. At the time I was in Norway, leading a seminar for farmers working with biodynamics. While I was meditating with the group from the seminar in the wonderful church at Stange, a mighty crown appeared in the space before me. It was magnificently worked in baroque style and richly sown with jewels. It seemed to me that I was being shown the Kingdom of Heaven, and I gazed on it with amazed enjoyment. Suddenly the crown burst into flames. Its splendor was devoured by the fire and soon there was nothing left of it. For the longest time I found myself staring at total emptiness. Finally I asked myself if there could never be a new beginning. At that moment a little bird emerged from the nothingness and at once flew off into the heavens.

I am not so well versed in bird lore that I could immediately determine what sort of bird it was. Fortunately I found it again in the Venetian painting by Giambattista Tiepolo, already mentioned. As described above, the bird sits on an iron rod opposite the Queen of Heaven, to mirror the perfection of the Goddess. The image gave me the opportunity to ask an ornithologist about the little bird. In his opinion it is a European goldfinch. (Translator's note for non-birders: The European goldfinch differs from the American in being multihued).

The bird expert also told me that there is a Slovenian folktale that God provided all the animals, one after the other, with the colors corresponding to them. The goldfinch came last in the line of animals and not enough was left of any color to give it a single hue. So God used what was left of all the colors to give the goldfinch its varied plumage.

Do we not again encounter here that somersault, so often mentioned, whereby the last becomes the first? The heavenly crown that symbolizes the very highest was transformed through the flames into the tiny goldfinch, last in the line of animals.

If it is so simple to change Earth and humanity, how does the transition work in practice? Since the current processes of epochal change are now fully developed, one should assume that we men and women are today leading two different but parallel lives. By day we are still fettered to the highs and lows of the "old" world. By night we already enjoy the "new" Earth and are gradually attuning ourselves to its legalities.

Now, in accordance with the principle of the somersault, our nighttime space should become our conscious space in daytime. As the changes proceed, we should learn to shift the focus of our attention from the old world to the center of our former nighttime space, which has now evolved to be the future space of life and light. Once this is done and people are changed, there can begin a long-term process whereby their individual aspects can emerge from the darkness of the unconscious. At the same time, the bonds that tie us to the old world structure will be released.

This may sound quite convincing, but when people who are not convinced of the Earth changes look around them, they see quite another picture. Practically no one among all the billions of human beings appears to be interested in what the Earth wants or in what direction the spiritual world is seeking to inspire us to move. The bonds that tie us to the old world structure are just about one hundred percent firm.

Despite this less than optimistic assessment, I have a firm intuition that the situation is not hopeless. I even had a lesson teach me that I am too easily influenced by outside circumstances. This was contained in a rather upsetting dream that I had on October 23, 2002.

In the dream, I am gathered with a large number of men and women, waiting to have a conversation with a spiritual master and teacher. The guru is constantly busy with people who come up to him with all sorts of questions on which they would like to have his advice. I am persistently trying to signal the master that in my case I am not concerned about personal matters but have urgent questions about the fate of the world. My efforts get me nowhere. Suddenly a woman puts a little stool covered with soft animal skin in front of the master and sits down on it. It is plain that she has a thousand questions to put to the guru. With her sitting there comfortably as if in her own home, I see no chance of my requests getting a turn in the foreseeable future. I feel I have no choice but to leave the place and go back to my daily tasks. While returning to my usual workplace, I chew on some slices of dried apple. They are so extremely dry and hard that I have to munch continually and dare not swallow them.

The first two dream images address the circumstances about which I was complaining above. Seekers of the truth must decide whether to be one of a group or to be self-sufficient. One may bind oneself spiritually to any of a variety of gurus or religious institutions whose view of the future offers deep and far-reaching certainty. This type of decision is based on a relationship that is charged with feeling and is usually characterized as loving dedication. In the dream, the animal skin on which the disciple was sitting points to an unconscious need for the emotional protection provided by supposed spiritual leadership in face of a world full of uncertainty and the processes of constant change.

Alternatively, one may choose the path of intellectual self-sufficiency. Then one is certainly not dependent on a guru, but the problems are no less. In the dream they are symbolized by my relentless chewing. This chewing never satisfies. Things are discussed, thought over and planned, but from the perspective of the approaching change, one advances not one step.

Back in the dream, I am led to a place in the precise middle of the above two extremes. A thunderous voice rings out, "It is time for the bells to sound." The bells are hanging in a long row on the wall and I know it is I who should ring them. But I am astonished to find that there are no bells among them suitable for sounding an alarm; these are small, finely wrought silver bells of different sizes and designs. I do not understand how these fine bells can have any meaningful effect. If any such silver bell is rung, who will even be able to hear it? So I am not prepared to do what that thundering voice has bid me. Instead, my first thought is that perhaps the time is not yet ripe to sound the bells; and secondly, that in any case people would find the sound unpleasant.

The second part of the dream points to a new and quite fantastic possibility that a person may neither have to submit to the dependence of a guru relationship nor be held caught in the nets of his or her own intellect. Quite clearly, this is the first time such an opportunity has become feasible for us humans, and it is a consequence of the processes of change and the inpouring of divine grace connected with them. Unfortunately, this method for solving our problems, symbolized by the silver bells, is so absolutely new and therefore apparently unthinkable that at first I had to reject it.

May I ask the reader for assistance? Let us together seek to answer the question: What wonderful things are now being offered us, as to the nature of which we have (still) no idea? The great numbers of bells that I have seen on the walls suggest that each of us may find our own variant of the solution, although they are all based on one and the same principle.

I should like to begin our group meditation with an exercise to open the heart center:

- Imagine that a wooden gate is standing in front of your midmost heart. The gate has wings on either side and is closed.
- Next you see the two wings slowly opening, and the light of your heart begins to shine.
- The wings open wider and wider until they come together at the back of your heart.
- From the united wings comes a strong root through which

your heart center is implanted in the heart of your cosmic twin behind your back. Your heart is open and its beams stream out to every side.
- You keep yourself firmly rooted in the heart realm of your cosmic double and apply yourself to our question: What unexpected possibility for change is being offered us, of which we had no inkling until now?

My own answer to this question was received in many roundabout ways over the course of the following month of November. It began with an invitation to fly to Ecuador to prepare a lithopuncture project in Quito, the capital. My flight went through Madrid, Spain, but on the way there, when we were directly over Barcelona, the computer system at the Madrid airport caught fire and our plane had to make an intermediate landing in Barcelona. In consequence I missed the connection to Quito.

I had to resign myself to my fate and wait for 24 hours in the airport hotel at Madrid. To fill in the time, I began considering the meaning of the above dream, which I had had 14 days earlier. However, I made no progress with it. I was given the key in a revelation that I experienced through another dream that same night.

In this dream I am standing in front of a high, broad stage. When I raise my eyes to see what is happening on the stage, I am thunderstruck. A dark-skinned giant of a woman is there, dancing. Never have I seen such power and beauty. The woman is completely naked and dances without moving from the spot. Yet somehow she moves every muscle in her body so that they all glide to the side in one direction. She makes a quick turn and now her muscles are moving in the opposite sense. Her dance is like the countless waves of a mighty sea, driven in one direction by the wind.

Rooted to the spot, I stare at the copper-colored Goddess, whose dance is like the rushing waters of a thousand oceans. Then I see the figure she is holding in her left hand: it is the Christ Child! It is borne in upon me that I have the privilege of looking on the divine

*Above me danced the Virgin Goddess. Gradually the features of her face took on a human character.*

Virgin. Quite obviously, she is going to show me something important. Now I can see that the Christ Child on her arm is put together from two boyish figures who are sitting so tightly together that it seems they should complement one another. But one child is pure and perfect; and the other is laden with crucifixes and other instruments of torture.

In fact, Christian iconography does include representations of two Jesus boys in one or other of these different roles. An example is to be found at the high altar in Constance Cathedral in Germany. High above us, on the right-hand side of the altar, stands the Blessed Virgin Mary holding a perfect boy who is raising his hand in blessing. On the opposite side, one can see the Blessed Joseph, who is bearing an identical boy on his arm, except that this one carries a cross.

Here we see the juxtaposition of two of the archetypal images that we explored in the first part of this book. One portrays the perfection of the inner child, who represents eternity within us. In the other we are looking at our cosmic double, who mirrors the memory of our inner development and is therefore obliged to take on our shadow side—until we understand how to transform the shadow in the light.

Back to the dream: Scarcely have I become aware of the two fold appearance of the Christ Child in the Goddess' hand than my eyes are drawn to her face. It is no longer clearly recognizable but is changing quite remarkably. These continuing changes begin with strange, mask-like facial features that have an exalted cosmic quality such as one sees in representations of the Goddesses of ancient cultures. No individuality is to be seen there, but they portray the All-Embracing, which includes the most radiantly lovely and the most hateful. Next follows a second phase where the features become ever more human. Through this she gains an individuality that is intertwined with a deep power of love. The Goddess becomes completely human—a concept hard to grasp.

Did God not make man—according to the Bible—in his own

image? But afterwards, so the evangelists tell us, that same God was born on Earth to incarnate in human form. Here, once again, we find the process of inversion, the somersault.

I see a similar process happening in the facial features of the divine Virgin on the world stage above me. As her cosmic grandeur is transcended and the human traits flow in, her presence becomes ever more meaningful, fuller of love and more possible to embrace.

No sooner have I become familiar with the language on the face of the Cosmic Woman than my attention is drawn to the musculature of her body. Amazed, I notice that in certain places her magnificent womanly body is temporarily taking on masculine characteristics. I see there a molded musculature that otherwise one only sees and admires in star athletes.

Afterwards, pondering the meaning of this dream image, I formed the opinion that it addresses the problem of patriarchal supremacy, which is ever present in our culture. But the longer I gazed on the rhythmic appearance of the male musculature in the course of her dancing, the more natural did it seem to me. Like the back of a whale that emerges from the wave-rippled surface of the ocean and at once vanishes, the masculine muscles continually surfaced here and there on the upper body of the divine, dancing Virgin and disappeared just as quickly.

Obviously, this symbolizes a unique partner relationship between the two genders of the divine, similar to that between the two polarities of the universe. To put it briefly: The masculine principle is present within the feminine aspect of the Godhead, just as the feminine principle is present within the masculine aspect of God. In the framework of Christian cosmology, the corresponding archetypal image shows Mary and Jesus both sitting on the heavenly throne, and Christ, who is already wearing a crown, putting a crown on Mary's head—one that is identical to his own.

My last three perceptions of the dancing "giantess," taken together, may be seen as initiations into the secret process whereby the Goddess becomes human: First, the divine Virgin takes on a body

*An architectural boss in the Gothic cathedral at Marburg on the River Lahn in Germany demonstrates the partner relationship between the two genders of the divine.*

that is definitely feminine but also includes the masculine principle; second, she descends and comes close to having a human form; third, she exhibits her son not only in his exalted cosmic role but also burdened with the shadow which, because of his charitable love and of his own free will, he has taken on himself.

In my last published book, *The Daughter of Gaia*, I had already begun to talk prophetically of the Feminine Redeemer. That was no simple decision for me, for I feel myself bound to the principle of human self-responsibility, and the idea that anyone should come from outside to rescue us from our entanglements was alien to me. However, I can no longer ignore the increasing signs that in fact a secret, cosmic redemptive plan has been set in motion.

The images in the dream make it plain that the matter is in fact about help "from above." The Virgin's dance took place on a stage above my head. For the whole time I had to look up to a plane of existence that lay a mighty step higher than the plane on which we stand here on Earth.

I was given no indication that the swinging, vibrant Goddess would ever leave her high spiritual plane and climb down below. That is certainly what one would expect from a redeemer or redemptress. Nevertheless, I was still holding onto the thought of a redemptress appearing. The matter was clarified for me as the dream of the dancing Goddess continued:

My attention is again drawn to the constantly repeating dance movements of the divine Virgin. I look with fresh enthusiasm on the coppery waves of her dancing muscles that—inspired by a powerful and inaudible rhythm—roll from left to right and back from right to left. Suddenly I notice that there is something askew about the muscles of the upper thigh. They are always lagging, as if unwilling to go along with the dance's changes of direction and reversals of motion. When the waves in the dancing body have already reversed their motion and, for example, are running from left to right, the waves of the muscles in the area of the upper thigh are still running from right to left for a few moments longer. The harmony of the dance movements in the whole body is only restored when the waves of the upper thigh muscles have also reversed, and then it only lasts for a short while. For then comes the next change of direction and the game begins again. Here, I was shown something remarkable. Every time the two movements begin their contrary motion, the Goddess' upper thighs become as stiff as two steel pipes. Their rigidity stops the wave-like movements. They start again when the two dance motions are harmonized for a moment.

This last part of the dream carried the message that the Redemptress wanted to show me a way in which she can be active in the alien earthly world without having to climb down off her "stage." It was a solution worthy of Solomon, one that preserved the independence of humanity and still put divine help and redemption within reach.

The Goddess is ready to start the vibrations on the Earth's surface dancing in such fashion as will draw the calcified and alienated vibrational patterns of humanity irresistibly into the vortex of regeneration and harmonization. We humans still resist the process of inner change, but this will enable it to be set in motion—whether we want it or not. However, the process itself cannot be carried to a successful conclusion by anyone other than the person concerned.

The wisdom of this solution lies in this: that circumstances will be created through which, in a redemptive sense, people can receive essential help. In general, however, to be able to enjoy the fruits of redemption, every person must cooperate creatively in his or her personal change—for example, through the exercises that are listed in the Appendix to this book.

Back in the dream, after I have realized, helped by my intuition, what a powerful gift the divine Virgin is preparing for humanity, I am so delighted and beside myself that the cry, "Mother!" comes ringing out from the depths of my heart. The voice of gratitude and devotion is so loud within me that I awake from sleep with a start.

The dream has answered many questions without, however, attacking the essential problem: How can people encapsulated in their own alienation be drawn concretely into the vortex of change? How can such a process take practical effect if the Godhead does not appear directly among human beings to inspire them to fundamental change?

The answer to these questions disclosed itself gradually during the following months as I made myself busy in two countries that are part of the great watery landscape of the Amazon: Ecuador and Brazil.

### Preparations for the Next Phase of Change

Quito, the capital city of Ecuador, lies on an elevated plain between two snow-covered Andean mountain chains. Some of the mountains are active volcanoes. The special power of the place lies in its situation—built along a mighty channel of light that, like an airy dragon, runs high above it in the atmosphere. It is through this channel of energy that the two great oceans, the Pacific and the Atlantic, exchange their powers.

The oceans are mirrors of the archetypal forces of the Earth's surface, comparable among humans to the space behind the back. Stores of energy are constantly interchanged and balanced one with the other through such channels. Through them the oceans give each other mutual support to maintain the fullness of their life forces and inner stability. It is my feeling that there is a twin channel somewhat to the north that touches the Mexican peninsula of Yucatan and the American peninsula of Florida.

My discovery of the light channel above Quito was confirmed by the presence, lined up along the channel, of some of the most important sacred sites of the pre-Columbian culture. Among these holy places there is a group of 15 step pyramids in Pichincha that must have had an especially exalted meaning. My perception is that these pyramids were built with a view to draw down part of the energy and let it flow along the Earth's surface. The formerly significant ritual center of Quito was thereby revived. The group of pyramids that brought this "bypass" into being has outlasted the conquest of the surrounding country, first by the Incas and then by the Spaniards, because they were covered with earth and plants and so resemble a natural hilly landscape.

The most critical of the former ritual sites to which the "bypass" leads is an interesting hill, now located in the middle of the Quito cityscape. Its shape resembles a giant step pyramid. The Spaniards jokingly call it Panecillo, meaning "Muffin." Its original name, however, was Chungoloma, "Hill of the Heart." The heart center that it accommodates is powerful.

When I let myself sink into the secret of the hill, on which a gigantic statue of the Virgin of Quito is now standing, I became aware of a very strong connection, supported by the archetypal forces of the transoceanic light channel, between the Hill of the Heart and the watery landscape of the Amazon Basin. It runs towards the east at a right angle to the light channel so that it touches a broad and water-rich mountain called Ilalo.

The meaning of the connection with Amazonia remained hidden from me until I arrived a few days later on the other side of the continent, on the Atlantic coast of Brazil. In fact, it was on the flight to Rio de Janeiro, which led me in an arc through Santiago in Chile over the watery Amazonian lowlands, that I was led to some essential insights.

One can perceive the characteristics of sites and landscapes very distinctly when one flies in an airplane high above the countries concerned. Of course, on the way one must attune to the resonance of the respective vibrational structures. In this case I also had a map of South America on my knees during the flight and was able to make my perceptions concrete with its help.

Some years before, when I was busy in Buenos Aires, I had realized that one wrongs the South American continent if, following the viewpoint of its white conquerors, one looks at it from top to bottom. To see the twin character of the American continent aright, one must view it from a Central American base.

Looking upwards from the standpoint of Central America, the

South American continent resembles a pregnant Mother Goddess. The mountain chain of the Andes and the Cordilleras respectively represent her spine; Tierra del Fuego at the top her crown. The giant, watery Amazonian lowlands, surrounded by various mountain chains, portray the shape of her "pregnant belly."

This gigantic watery belly, where something new is being created, is related to the revelation of the divine Virgin in Madrid. I could sense that all the energy channels that lead from various directions to the "pregnant belly" of Amazonia are, at this very moment, at the point of being revitalized. I have already mentioned the link between Quito and Amazonia in this connection. A second energy channel runs from the Pacific Ocean across Lake Titicaca towards Amazonia. Furthermore, there is a third sequence of chakras that has twin origins in Sao Paulo and Rio de Janeiro, and then stretches out to Amazonia through Brasilia, the capital city.

In the next few days I had the good fortune to work with groups engaged in city-wide healing in Sao Paulo and Rio de Janeiro. Also included in this healing work was a landscape in the southern portion of the state of Minas Gerais where the two energy channels from Sao Paulo and Rio de Janeiro flow together. This landscape is called Circuito das Aguas, "The Wreath of Waters," because of its countless mineral springs. Some years earlier, I had had the privilege of erecting a lithopuncture system for Circuito das Aguas.

Next, I had a dream that indicated what was happening through the activation of the watery Amazonian belly through the three energy channels:

The emotional power field of Amazonia is shown me in the form of a mighty lake. Suddenly, I notice that the color of the water is beginning to change. It becomes a shining green, as if the water were lit from within. I have never before seen the like of it in any lake. Next, the colors on the lake's surface even begin to differentiate. Beautiful, strong and harmonious patterns come into being and transform themselves into rhythmic movements. They run vigorously, first to the left, then to the right, making a clear connection with

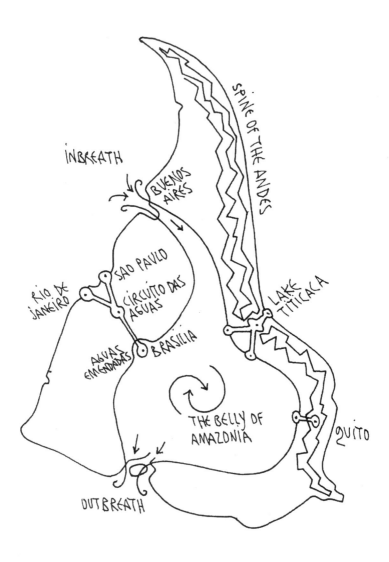

*The South American continent with Amazonia, the "pregnant belly." Also shown are the three light channels that nourish Amazonia with their spiritual-emotional power.*

the wave-like dance of the divine Virgin from Madrid. I look around me, seeking the source of the impulse that is causing this transformation of the lake. All the rivers of the surrounding countryside are emptying into the lake; they carry on their surface the characteristic pattern that eventually gets changed into the above mentioned wave-like movements by the "emotional water" of the great lake. As I watch, it occurs to me that one of the rivers is flowing towards the lake from my direction.

Transposed into everyday language, the dream's message tells us that a mighty force field is being created in the watery belly of Amazonia, one that vibrates in harmony with the cosmic dance of the epoch-making change. The above-mentioned energy channels are activating this field and defining its quality. Its unique dynamic and power will attract the feeling-fields of worldwide humanity so strongly as to raise them out of their old patterns of ideation and imagination. This will give us a new opportunity to make good on our threatening backlog as we too engage in the process of epochal change.

According to my observations, such magnetically attractive emotional-cosmic force fields are also situated in the Lagoon at Venice and in the middle of Asia, more precisely between Alma-Ata and Lake Balkash in Kazakhstan.

**The Future Scene of Change: The Spiritual-Emotional Plane**
On the day that I was planning to leave South America, I got the news that a Brazilian shaman, whom I did not know personally, had dreamt of me the previous night. By good fortune, he had told the dream to somebody who had just been to my workshop in Sao Paulo—and so could recognize that its message related to my researches on the light channels that are reviving the watery belly of Amazonia in preparation for its new role.

In this dream, we are traveling together towards Chile. When we get to the country's frontier, we see slaughtered animals hanging head downwards from a rope. There is nothing tragic in this, for the animals have been ceremonially sacrificed as part of a shamanic ritual. Their blood, dripping to the Earth, acts so beneficially upon the groundwater that perfectly pure water arises from the soil.

In a second sequence, we are in a wood and all around us we see animals stamping on the ground. And wherever they tread, we see crystal clear water beginning to well out of the Earth.

In a third sequence, we are both running through the woods because we have been told that people should sweat as much as possible. Each drop of sweat that falls to the Earth has an effect like the drops of blood from the sacrificed animals—crystal clear water arises from the ground.

This dream, which reached me by a detour through an unknown shaman, was intended to direct my attention away from the planetary level back to the human process within our personal Holon. It was not by chance that a shaman was chosen to be the dreamer. Shamans guard the wisdom of the cosmic origin of animals. The sacrificed animals in the dream are hanging down from heaven because the various species of animal represent an incarnation of different cosmic archetypes. In this connection one may think of the zodiac, which represents twelve archetypes (Translator's Note: Our word "zodiac" is *"Tierkreis"* in German. *Tier*=animal; *Kreis*=circle).

Animals are mediators of these archetypal images, bringing their qualities from cosmic space into the realities of the earthly Holon. The Earth's miraculous response to their drops of blood can best be understood as pointing to this role. The archetypal cosmic images, working into the earthly plane, cause Earth to bring forth wonderful powers of healing, purification and reinvigoration. This fact, today unfortunately forgotten, is confirmed in the second dream, which emphasizes that this role is accomplished through the pure being of the living animal and in no way through death or sacrifice.

My experience with the animal world tells me that animals are completely permeated by a deep cosmic-emotional purity such as is scarcely ever found in the human being. The emotional purity of animals is of an absolutely fantastic, crystalline clarity. In this respect the human being could become a disciple of the beasts. Here I am thinking not so much of domesticated animals that are permanently exposed to human influence. They possess other qualities. I am referring to simple animals living in the wild, like birds, fish, dragonflies or deer. A young animal, for example a lamb or fawn, would also qualify.

We are dealing here with a plane of existence that is almost completely lacking in our modern consciousness: *the spiritual-emotional plane*. It can best be perceived if the animals are invited to be mediators. I call the following exercise, "The Animals' Personal Gift:"

- Using your imagination, invite the animal you have chosen into your Holon. There you let it swim, chirp or graze—or simply be.
- Ensure that the emotional quality embodied in the animal spreads through all your personal space—to dissolve all the illusions, chaotic circumstances or unclean places that reside in crevices within the sphere of your emotional power. Simply let it happen and follow the process with awe and amazement.
- Then give thanks and let the animal go free.

Let us return to the unknown shaman's dream and its third image, where the drops of animal blood are replaced by drops of human sweat. This brings to mind the dream I described many pages previously, about the little silver bells that should begin to ring during the present phase of Earth changes. Sweat is a fluid that is produced from within a person by his or her exertions, and the delicate silver bells only have meaning if they ring within a person. Their sound is too weak to have any effect on the outside.

This means that a capability similar to that of animals exists in human beings. There are countless human cultures that in their own way have translated the images of cosmic archetypes into rituals, myths and religions, and these bear witness that such a capability is not foreign to us. It only seems foreign because, unfortunately, we have learnt too well how to transpose this capability into collective formats. However, the two dreams point to the possibility that the same quality would still be effective if it were to be used by an individual. It follows that one should learn to embody the cosmic archetypes through one's world of feeling.

This need points to an approaching task, exemplified by the dancing Goddess from Madrid, who presented three of the basic cosmic archetypes: first, the image of the inner child; second, the secret of individuation that I read in the changing features of her face; and third, the balance between female and male. These archetypes were transposed into watery patterns of movement in the course of her wave-like dance.

This means that our incorporation of images of cosmic archetypes into our world of feeling is key to the new phase of earthly and human change. In regard to Earth as a whole, the incorporation of cosmic archetypes is accomplished through the change of certain watery landscapes, as represented by Amazonia, for example. What is preventing us humans from giving space in our personal lives to this exciting task and energetically cooperating? What is preventing us from reviving the images of spiritual archetypes in the mirror of the emotional plane?

My dream of February 22, 2003, offers an answer to this question. In this dream a group of us are visiting a Tibetan temple that is like a place of initiation. We arrive in a car. Before getting out, I take off my shoes because I suspect that we shall be performing rituals where those attending must be barefoot. First, we have to pass through various buildings that represent individual steps in the initiation process. I am well versed in the performance of ritual ceremonies, so I take rather longer to pass through the buildings than do the other members of the group. They have already gone on when I leave the little temple and come out into a broad courtyard. In the meantime it has grown dark. In the shimmering moonlight I perceive that the courtyard in front of me is rather wet and marshy.

Having left my shoes in the car, I realize I could get wet if I cross the courtyard. This would not be so bad if I were really barefoot. However, I left my socks on when I took off my shoes. Now I am standing there in black, knee-high socks, and I know that they will get wet if I proceed further on the path of initiation. The only way I can keep my feet dry is to turn round and fetch my shoes from the car. On the way back there, I think of another solution. I could renounce the path of initiation and go to the temple's dining room, and there eat and drink and wait until my colleagues have completed their initiation. I decide for this last option. Now I am in the dining room and eating and drinking my fill. Much time has passed, but the rest of the group is nowhere to be seen. This unpleasant situation finally convinces me to go back and take the first option. Although it has meanwhile become pitch dark, I start towards the place where I left my car, meaning to put on my shoes and then go to look for the rest of the group. But in the meantime my car and my shoes have vanished.

To go barefoot on the Earth means that one is keeping a free and direct relationship with earth and nature. Shoes are a symbol of humanity's culturally conditioned and alienated relationship with earth and nature. To take off one's shoes signifies a wish to regain one's original connection with earth and nature.

In today's world many people are inspired by the same attitude as I displayed in the dream when I took off my shoes. It may be that the general change in Earth and humanity is playing a role. The process so far has led us through various phases, which I experienced in the dream as stages of initiation.

Now we have arrived at the phase involving watery power fields and the emotional sort of archetypal quality. It is clear that the symbolic act off taking off one's shoes is insufficient to enable one to pursue the epochal process of change any further. "In order not to get wet," one must now take off one's knee socks.

There seems to be a far-reaching problem bound up with this, because I am willing to abandon the very promising path of initiation rather than take off my socks. The two alternative options that I choose, one after the other, have shown themselves to be unsuitable. Yet the idea of simply taking off the socks and going barefoot into the wet never seems to enter my head.

In clarification, it has to be said that socks are in no way shoes. Here we are not looking at the symbol of any tangible relationship with earth and nature, but at a much more transparent level of relationship. If one decides to walk in water, one's feet will get wet whether or not one is wearing socks. The transparency of a sock's fabric points to an emotional relationship that has its home midway between spirit and matter. I prefer to speak of it as a spiritual-emotional relationship.

The symbol of the sock has a second aspect that plays a role in this. The dream is specific about portraying knee-high socks. Knees are a traditional symbol of the soul. The inner meaning here is that the soul is not so deeply entered into matter as is the body. The body touches the Earth's surface through the soles of the feet. In contrast, the soul extends downwards no further than the knees. Thus, a kneeling person is really standing on the feet of the soul. That is why we kneel down to pray.

Following this logic, knee-high socks, which cover the zone between the Earth's floor and the knees, represent the relationship between a person's spiritual soul and the surrounding manifested world. This is a relationship that we can sense, depending on the degree of our spiritual presence in the here and now. If one lets one's soul-powers stay at knee level, one lose one's soul's ability to work subtly into everyday life. One's own spiritual treasures are ungrounded and begin to float up into the wide blue yonder. On the other hand, if one stays fixed too firmly in one's shoes, the voice of the soul tends to become mute because it has lost the subtle connective tissue, symbolized by the knee-high socks in my dream. Communication with the soul is broken.

I should emphasize that this time we are not concerned with a spiritual relationship with the soul, as was the case in the exercises relating to the inner child. Here we are concerned with an emotional relationship through which the archetypal images of the soul become realized in daily life. If the salt of the soul is not mixed with one's daily bread, life becomes empty and arid—as in my dream, when I withdrew from the path of initiation.

To experience the spiritual-emotional quality of the precious subtle tissue connecting the soul with the reality of everyday, I propose the following exercise:

- Kneel on the ground.
- I am sure you have seen plants stretching out almost transparent roots, not into the Earth, but in water. Allow similar transparent roots to grow from your knees into the depths beneath you. As you do so, imagine that the Earth is not composed of a firm and rocky crust, but of water.
- Give every single one of those fine roots a different rainbow color. Let them penetrate the watery depths of the Earth and follow their wave-like streaming.
- Finally, stand up and lift the roots without letting them tear. The goal of the exercise is accomplished if you succeed in keeping the emotional roots of the soul intact whatever you are doing.

It seems impossible to say, unambiguously, where the process of change will next lead us. There are various developmental threads that focus on the regeneration of the emotional plane. The last indication I received before finishing this manuscript was contained in a dream I had on May 2, 2003. In this dream I am carrying the upper part of a plaster figure of an ancient Goddess across a courtyard when all of a sudden her hand comes alive and strokes my right cheek with a touch as light as a breath.

The dream leads us back to the dancing Goddess of Madrid—and the Cosmic Mother's quite amazing ability to reinvigorate our world-space, which is apparently so deeply asleep. Mighty conjunc-

*The deep rootedness of the soul—an exercise.*

tions of purified emotional fields come into being, all permeated with the power of cosmic archetypes like those described in relation to Amazonia.

These spiritualized emotional fields are spread out worldwide from their sources and are available to human and other beings struggling with their processes of change. If one attunes to the special vibrations of these fields, one can invoke the desired help, support or inspiration. It is also possible for an unconscious attunement to simply arise for an individual or group from inner impulses in a given moment. Then the world suddenly becomes lovelier, relationships deeper, and one would like to shout aloud for sheer happiness.

So far I have identified three such spiritual-emotional (watery) fields; to be precise, the emotional fields of:

Healing,
Relationships between related souls,
Hedonism.

But how can an individual consciously attune to these fields?

One should build the contact with the corresponding field either on one's back or at the back of one's head; this is what I propose:

- When kneeling, standing or lying down, imagine that you are simultaneously bending backwards and raising your feet up behind you to such a degree that the back of your head touches the soles of your feet. This contact rounds you off, and, at the point where the contact takes place, it makes it possible to connect with the desired spiritual-emotional field.
- Then let the image go, and for a while breathe the desired quality into your body and your Holon respectively. Rejoice for the gift you have been given.

CHAPTER EIGHT

# The Present State of Affairs

**Problems with the Exaggerated Yang**
WHEN ONE HAS ONCE UNDERSTOOD that the world situation is to be viewed as a continuing message—at any rate since the Earth changes began—it is not hard to interpret the situation as it presented itself in the spring of 2003. On the one hand we have the war in Iraq, where hundreds of thousands of soldiers have been fighting. They are equipped with weapons of the highest technical development. Death is sown with a hard hand and a heart ideologically brainwashed into silence. At the same time we must watch powerlessly as the tiny, deadly virus of an atypical pneumonia spreads fast as lightning across the globe. People die despite the no less elaborate "weapons" and high-tech apparatus developed by modern medicine.

In both cases we are observing the effects of an extremely exaggerated state of Yang. The first case illustrates our civilization's excessive masculine power, which goes hand in hand with the killing of fellow human beings and destruction of the environment, using instruments representing the highest technical "advances" that daily progress to new levels. The second case shows the helplessness of the swollen Yang when it is confronted by a microscopically small being out of Nature.

Over the past year, my dreams have indicated increasingly that we can soon expect a phase of Earth change that will touch on the enormous imbalance between the Yin and Yang forces in modern civilization. Actually, this book began by addressing this same theme, in the image where a man turns round and lifts his pure Yang—i.e., the boy in him—into his arms. The Yang was to return to its original source. However, it was my dream on February 14, 2003, that first made clear that the approaching equalization of Yin and Yang forces will be the focal point of the processes of the changing Earth.

In the dream I am being fed with gold coins. I hold my mouth wide open while the first gold coin is laid upon my tongue. The ceremony reminds me of the ritual when one takes the Host at the Christian Communion. It is terribly difficult for me to swallow the shining coin. How I wish for some water so that the coin can slide down more easily! As if looking at an X-ray, I watch the coin slip slowly and laboriously downwards in the direction of my stomach. The color yellow is a Yang color; the shining brilliance of gold represents the Yang power of the sun. Money, which rules and controls the entire world, is a symbol of masculine supremacy, i.e., of the Yang powers, in our civilization. My burning desire for a drink of water is, in contrast, a significant sign of its lack of feminine Yin powers. Also eating, viewed as the incorporation of organic nourishment from the natural world around us, is a Yin process. For gold to be eaten in ritual fashion symbolizes such supremacy of Yang as will lead us directly into cataclysm.

We should not overlook that it was on the personal plane that the dream demonstrated the dire effects of Yang superiority—and not on the international, collective plane, where it rampages currently. Does this mean that it is also in the realm of the individual that the solution is to be found?

How personally to participate in solving this top priority current world problem was shown me in a vision that I received at a seminar in Kempten, Germany, during an exercise to balance Yin and Yang. The exercise, in which a woman experiences her inner man and the man his inner woman, is detailed on page ___ in the first chapter.

When doing this exercise, I had until then always perceived the woman within me quietly looking in a backwards direction. But this time she sprang quickly out of my skin, turned herself around to face life and spread her arms wide in a protective gesture. Reacting, my Yang part withdrew thankfully into the backward space vacated by the woman, and there fainted.

The vision corresponds to the archetypal image of the Pieta where Mary, Mother of God, is holding her dead son Jesus in her lap. The woman grieves deeply, but she is strong enough to carry the fatally wounded man. The exaggerated power of Yang can only be slowly healed by experiencing the collapse of its superiority. However, the experience does not propel the man into final death, because at the dramatic moment his collaboration with the woman within him creates a protected space in which his fatal wounds can heal. The process of resurrection approaches.

For a man or woman, the new way of balancing Yin and Yang may be best illustrated by allowing one's own masculine aspect to withdraw within, there to embody the inner power. The Yang power then stands at your back like a golden yellow pillar to lend you the courage and inner clarity to change.

By contrast, the woman in you spreads to your front like an opening flower. Her Yin quality is made manifest in the colors green and blue. She is loving, competent and brave, and ready to step forward and face the challenges of the changing era.

Once again, we are looking at a somersault: what previously had been overly exalted is now withdrawn within and reversed. For this I recommend the following breathing exercise:

- *First Inbreath:* Draw the breath out of the width of the universe into the area of your larynx—the color is *crystal white*.
- *First Outbreath:* Breathe out and vertically down to the pelvic floor to build the pillar of Yang—the color is *golden yellow*.

*The modern Pieta: Our Yin supports the space into which the exaggerated Yang may withdraw and come to himself again.*

- *Second Inbreath:* Draw the breath from the depths of the Earth into the heart center above—the color is *green*.
- *Second Outbreath:* Breathe the power of Yin from the heart center out into the space in front of you—the color is *blue*.

To continue the exercise, begin again with the "First Inbreath."

**One More Warning**
This warning relates to the problem of the dwindling powers of life of which I have been made aware over the past year in connection with the progress of Earth change. The warning is best understood with the help of a dream that I had on January 20, 2003.

In the first part of the dream I am reminded that the powers of life are gradually being withdrawn from the "old" world structure. Earth will be concentrating more and more on the expansion of the newly forming etheric space. This proposition was discussed in the fourth chapter. This includes a process by which the vital powers are withdrawn from the plane of daily life where we and our current living processes are settled. This will be extremely challenging for us humans. And yet we must still manage with the "old" space for a while—until we have all learnt to breathe within the new living environment.

In the dream this unhappy situation is presented as a rental relationship that I have concluded with a "Guardian of Life." The agreement letting the space states that I will pay the slice of a golden apple by way of monthly rent. The apple stands for the fullness of life's power. I pay the rent several times without any problem, every time biting off a slice of apple with my strong teeth. Then I realize that the slices are becoming ever thinner. Finally, the apple slice is so meager that I am feeling deeply ashamed when I prepare to give it to the Guardian. It occurs to me that instead I could pay the rent with a material that has accumulated in my sacral area. This is in fact a repulsive ball such as often gathers in drainage filters and is composed of hair, filthy grease and other waste. In order to bring it into some sort of shape and use it as a means of payment, I stick the ball in my mouth and begin to chew on it. It has a nauseating taste and I am unsuccessful in getting the stiff, dead material to take on shape of any sort. I am wishing I had a little press to mold the material into some simple shape without having to put it in my mouth.

The dream tells us that we are approaching the point when the living power in the environment will have decreased to such an extent that it can no longer energize us. At a certain point in time, the Earth organism will no longer renew it. If we humans identify ourselves exclusively with the old world structure any longer, increasingly we will have to work with dead material.

One difficulty here is that we cannot even acknowledge the absence of life powers because, in general, we modern people lack the capacity to judge the inner quality of our food or our surroundings. The other difficulty is that it is hard for us to accept the catastrophic lack of the powers of life in our old, habitual etheric space because within and around us Nature is already vibrating to the frequencies of the new etheric space and so comes to us healthy and full of life. Her beauty and luxuriance will only increase in future.

But what use to us is the life power of the new etheric space if we are still at the mercy of the legalities of the old world structure and remain coupled to the power fields of the old etheric space? There is a danger that the human race will succumb to a lack of life's power in the midst of life's abundance. Such a tragedy cannot be allowed! What can we do to turn things around in time? My proposals are aimed in two directions.

First, we should continually exert ourselves to withdraw our attention from the old, mentally created world structure, for it is in the process of disappearing. Earth and Cosmos are no longer prepared to keep alive the old, already discarded spatial structure for the sake of our egocentric wants. Instead, our attention should continually be directed to the vibrating, dancing essence of life that is thereby drawn to meet us, and we should not be put off by the manner of its appearance. If our attention is directed from within by the heart's power, it will always find the right goal. *One is present exactly where one fixes one's their attention.*

At the same time, we should become accustomed to be continually severing ourselves from the old, vanishing reality. Certainly, we must take it as it is and tolerate it for as long as it takes for everyone to learn to connect to the new, life-filled reality. But we should not stay coupled to the old world, energetically, in thought or in feeling, any longer than is necessary. In a conscious and loving way we should keep our distance from the old world structure. We can use the following exercise for the purpose:

- Be present in your midmost heart. Sink down into the Silence.
- Imagine that your heart center resembles a golden sphere. This sphere is surrounded by several layers of crystal clear light. These layers are like a fine sieve.
- Using the power of your heart, push these sieve-like layers of light through your energetic and bodily structures, tackling them from all sides. Push them through your emotional fields. All the patterns and forces that belong to the old world are thus sifted from your cells and power fields.
- Ask that all your ties and dependencies that relate to the vanishing old world structure should be withdrawn from your body. In this way, all the "old junk" is brought so far out into the open that you can finally push it from your Holon.
- Then gather the "junk" into a separate ball of energy that is clearly apart from the sphere of your Holon. Give thanks for the teachings that these ties, now severed, have imparted to you. However, express very clearly that henceforth you will have no more to do with them.
- Then imagine that this ball is guided very quickly out of the remains of the old world into the realm of change that lies in the distance. There it dissolves itself in light.
- Be thankful that you are conscious of the true shape of wholeness.

A dream I had on February 15, 2003, indicated another way to prevent the dwindling power of life from carrying one away with it:

I am visiting a great art exhibition. Surprising even myself, I suddenly decide to steal a drawing. The plan seems quite illogical, since I am myself a draughtsman. Apart from this, the drawing's style resembles my own. I quickly take the drawing out of its frame, which is square-shaped; I hide it behind an ornamental plant. Then I roll up the drawing and stick it in my left shoe so that the rolled-up drawing lies against my leg and reaches to my knee. I draw the trouser leg over it to hide it. Although the theft has been smartly executed, I am afraid of being caught.

It was nearly three months before I understood that there are precise instructions hidden in the dream how to go about regenerating the old reality without exposing oneself to the questioning gaze of those around you.

These instructions have allowed themselves to be transferred into an exercise, as follows:

- *The drawing is taken out of its frame:* With the help of the imagination, you should lift whatever asks for renewal out of the framework of reality.
- *The drawing is rolled up:* The relevant facts, or the situation in question, should be handled in the context of the three light channels. The heart channel can be recommended in all cases. In certain cases the forehead and lumbar channels may also be appropriate. The lumbar channel is indicated if the treatment has to do with a lack of the source powers of life, the forehead channel if a spiritual blessing is needed.
- *The roll reaches from the soles of the foot to the knee:* This image tells us how it is possible, with the help of one of the three light channels, to renew the dwindling power of food, for example. The three channels represent the bridge that connects the limitless vital powers of the spiritual soul to the reality of everyday. As stated above, the knee is a symbol of the

eternal soul. The soles of the feet represent the point of contact with the ground.
- *The trouser leg is drawn up over it:* This action can be carried out almost unnoticed. For example, when the meal lies ready before you on the table, briefly shut your eyes and ask for the purification and reinvigoration of the food. In your imagination, take the essence of the meal briefly into your mouth and so bring it into the region where it can be blessed by the forehead channel that runs just above. Then let it sink into your heart space to regenerate it and, finally, still deeper into the pelvis for it to be strengthened with life's original powers through the lumbar channel. Then reunite the essence of the food with the meal on the plate in front of you and give thanks. The exercise can be adjusted to the exigencies of the moment and switched around as you wish. You will find the basic text in the Appendix.

A really stormy period lies ahead. Let us mutually wish ourselves good traveling through the tumults of change.

I hope to have provided you with the tools necessary to stand up to the challenges of this epochal renewal. If, in spite of this, something seems to run awry, there always remains the one unsurpassed recourse: You fall to your knees, connect yourself to the spark of light in your midmost heart, and you ask. Remain faithful to the Mother, whom we call Earth.

**Release from One's Own Entanglements with the Dark Powers**
The summer of 2003 brought long-lasting drought and unbelievable heat. Fire raged in various parts of the Earth. Parallel to this, in the middle of 2003 a certain theme was taken up anew, one which I have discussed several times in the book. It concerns a further deepening of the theme of the contrary powers. I have been made aware that these are not only powers that work against the process of change, but also powers that have a burning interest in their own release.

In the fifth chapter, I told of the cunning strategy developed by the contrary powers to ensure their own survival amid the circumstances of the changing world structure. They initiate and encourage "wars for democracy and freedom." Or they call vociferously "for peace," feigning public benevolence to reinforce their supremacy on the worldly plane. Examples can be seen in the world events around us.

As described in that chapter, the contrary powers are not only busy with the gigantic accumulation of negatively charged emotional forces, which humanity excretes *en masse* when people allow their lives to be impregnated by fear, hopelessness and falsehood or other destructive thoughts and feelings. Also, because various individuals or power-hungry groups in the past—and probably it also happens in the present—have often communicated with the mass of destructive forces and harnessed them to their self-seeking ends, the contrary power has been able to acquire intelligence. Today it can pursue its own goals and struggle for its own survival.

In the middle of the year 2003, I was alarmed to find that the mega-shadow of the contrary power is habitually working to weaken the after-death path of discarnation that leads a soul to purification and re-acceptance into the spiritual world. For the purpose, it uses large numbers of souls that have succeeded in gaining the necessary knowledge. In the moments following death they exploit the accumulations of the so-called "lower astral"—accumulations of negatively charged emotional forces—and there hide themselves.

I should emphasize that this is not just a matter of soul sheaths left behind in Earth's astral realm by people who have suffered depressing or extremely traumatic deaths. Here we are looking at a soul's conscious attempt to avoid the divinely appointed path and keep itself in relative proximity to the material world after death. There, it cannot be controlled by the spiritual world and can pursue its own egocentric goals.

Within such souls there dwells an unshakeable conviction that they have not really died and have become the unique master of world circumstances. They try to guide individuals or even world events with their "messages" and "inspirations." Their effects on the world can be seen in the illogical twists in the path of world events, which in the end lead many people into harm's way. If individuals habitually act under their influence, they exhibit the most varied sorts of spiritual extremism, to which they hold very stubbornly.

In regard to this disagreeable theme of the contrary powers, my reaction is the same as most good-natured people. I want to know nothing of them. However, the experiences and messages of the past two years have been trying to direct my attention along the precise path that I do not want to travel. It seems as if the world soul is presently being offered the cosmic opportunity to expedite the release of that shadow aspect of the planet and humanity that we have just been discussing.

To give wings to my interest in the task, a coal-black Goddess has appeared to me several times in the last few weeks and signaled her readiness to examine her domain of the shadow in the change process if humanity is prepared to take up its share in the redemption process. I have perceived her to be a representative of the feminine aspect of the contrary powers, a particular aspect that has never lost its relationship to the divine. Redemption is finally being made possible by a somersault on the path of change.

As this book has repeatedly emphasized, if one looks on change as a personal challenge, it is possible to achieve its goals however overwhelming the task may seem. If the problems in the realm of the human microcosm are addressed and healing introduced in that area, the result will be an unstoppable healing process on the plane of the planetary Holon.

In fact, there is a shadow phenomenon within the personal Holon that resembles the problem of souls gone astray in the astral world (the feeling world of Earth). Some psychologists designate this phenomenon by the term, "Pact with the Devil." The problem was first presented to me in a dream on July 13, 2003.

In the dream, I am bicycling with my wife on the road to

Jerusalem. During the journey I remember that I have left my money wallet at home. Without money, we cannot reach our goal. What can we do? Finally it is decided that I will go back and fetch my black money wallet. My wife will continue the journey and I will catch up with her later.

In the dream's next image I am once again on my bicycle on the way to Jerusalem. Provided with a sufficiency of money, I am hurrying to catch up with my wife. Night catches me near an abandoned bus. I climb in and sit down in front, in the driver's seat, to have a little sleep and renew my strength. I wake up (in the dream!) at the very moment that the bus is being attached to a private car and towed away in some unknown direction.

I am horrified. There I am, sitting in the driver's seat, but unable to steer the bus where I want to go because the thieves decide our direction. I try to sound the horn to alert the people along the road that the bus is being stolen, but the horn does not work. There are people standing on the edge of the road. What shall I do to make them aware of my hopeless situation? For a moment I breathe easier because among the crowd I see a group whom I take to be police officers. As we go by, however, it turns out that they only wearing costumes that resemble policemen's. I am in despair . . .

Transposed to our usual language, this means that at a certain point in time during a past life, there was some magnificent prize to be won that is imaged by the journey to Jerusalem. However, I did not have available to me the enormous strength needed to realize my desire. So I borrowed the necessary strength from the gigantic storehouse of the alienated powers—from Lucifer's gold pit. That is why I vanish from the first part of the dream—to fetch the black wallet full of money.

One does not need to feel ashamed about this or project a bad conscience for oneself prematurely. It is true that such loans were habitually used to strengthen personal power, in the course of which many of one's fellow human beings suffered harm in the end. However, the "Pact with the Devil" was often used to create mighty works of art or lead one's fellow humans into the inspiring experiences of certain advanced cultures.

However, my dream does point to an energetic problem that has its source precisely in such practices. The fateful coupling with the dark power has brought alien forces into our own human Holon. Through the consequent, often dubious, deeds, they were "frozen" into our force field and have remained frozen in their negatively charged state over long periods of time. Like dark stains, they influence our present moods and decisions from the abyss of the unconscious. Their effect is similar to that of other personal shadows, in that the individual ascent to the newly preparing plane of being cannot be accomplished if these dark stains are not released from the shining robe of our soul. The forces, once released, shall be purified and given back to whatever was the original source of their alienation.

In my dream on the following night I was expressly asked to follow the tragic effects of the past coupling with the "the elegant offer of the black banker" and see how they are working out currently in relation to the hoped-for Earth change.

In the dream, we are doing our national service as conscripts in the army. But just then the legislation changes (meaning, the Earth change is underway) and the army is duty bound to release us prematurely. The responsible officer is not at all in agreement. In his opinion, it is still his right always to draft us into military service if he so wishes. To prove his right to me, he shows me the black spiral rope secured to my foot, with which he holds me bound.

I become enraged and tell him that I have long ago paid back my debt through my works of art. I threaten to call a lawyer to protect my rights. At the same time I know that the black rope cannot easily be escaped.

One cannot free oneself from the black rope. One can only be released from it through the power of love. That power expresses itself through forgiveness and gratitude, given to the beings of the shadow for all that we have learnt through our wrong choices.

For this purpose I was given an exercise that relies on the potential of the newly broadened heart channel. One should not shy away from contact with the darkness, but rather feel protected, for actually it is cooperating in the release of the alienated portion of the world soul.

- Take yourself into the inner Silence, ground yourself, strengthen your protective sheath and center yourself.
- Be aware of the heart channel that runs horizontally to join the three aspects of your heart chakra: the cosmic heart behind your shoulders, the heart chakra itself and the outflow chakra in front of the breast.
- Build a sphere of light at the end of the heart channel in front of your breast. Decide whether you wish to work on the release of the conflicted soul portion within you, the alienated fractal of a specific place or a darkened fragment of the world soul.
- Gather behind your back the feelings, situations or symbols that relate to the alienated aspect of the soul substance on whose release you will be working. However, you can be open to whatever is standing in line for change.
- Allow the unhappy force that you have collected behind your back to flow through your heart channel into the sphere prepared for it. Accompany its movement with a prayer for its forgiveness and release.
- When the dark soul substance has finished its passage, or wishes it to be broken off, you should close the entry at the rear and completely empty out the channel into the sphere.
- Next you should imagine yourself taking the light sphere into your hands *and uncoupling it from your heart channel.*
- With the help of your imagination, you should now stimulate the process of change within the sphere. You can use colors for the purpose, e.g., violet, or work with a vortex of water, initiate a dance of vibrations within it, etc.

- When the change has got under way, you should call the cosmic messengers to you—the angels—and ask them to take over the light sphere and bear it to wherever the changing of the soul substance contained therein can be victoriously completed. Purified, it shall be given back to the place where it was alienated.
- Give thanks and experience the happiness of liberation.

**Centering Earth's Etheric Space**
The process of Earth change can best be understood in the following way. The all-pervading consciousness of the universe—alias the feminine-masculine Godhead—offers us, as the evolved ones of Earth, an opportunity to reverse our path towards human and ecological collapse. Within the context of Earth change, so-called, we are constantly being offered cosmic opportunities to resolve the accumulated problems in Earth's etheric space. We perceive them as various phases of the processes of change. Each phase offers a wonderful solution to the problems. But stop! These solutions will unfortunately remain hanging in limbo and never become real if we human beings do not perceive the possibilities offered us and at least try to anchor them in the reality of every day.

We can accomplish this through meditation, or by adopting a different attitude towards a particular aspect of life, or through activity that is attuned to Wholeness. However, I must emphasize that Earth change holds out no automatic promises. The only thing that's automatic is the further destruction one is facing! Recovery from that, however, is based on personal decision and individual creative initiative, or conscious group creativity.

This is why we human beings need to discuss the matter with one another, support each other with our perceptions of change and anchor them among us by holding meditations in common. Otherwise, we will again and again be offered wonderful opportunities . . . and they will expire and sink into forgetfulness.

I fear that is what happened to the gift of the unique constellation in the heavens on November 9, 2003. Standing in the sky on that day were nine planets in the form of a Star of David. I had flown to Scandinavia just the day before, so was well placed to perceive the preparations for a cosmic event. High in the heavens I saw how a golden network of inspiration was building to join the single focus points of the Christ power one with another. From Earth arose a graceful, silvery network of her powers, sweeping shadows in the change processes along with them. One could sense that a future-oriented quality was coming into being in the space between the two networks.

Have we done enough to ground what was offered us on 11/9 in our everyday reality? That is hard to evaluate; I can only say for myself that I had no idea what step to take so that I could embody the offering. In any case, the poles of the event's development were reversed at latest on November 24, 2003. I was made aware through dreams and spatial perceptions that there is a dramatic collapse approaching if we simply allow the gift of that unique constellation to pass us by. However, at the beginning of the new year I still had no idea of what that meant in practice.

It was on January 18 that disenchantment came upon me like a cold shower. It happened all of a sudden through some of us in the Life Network of Geomancy and Change getting into conversation and exchanging our experiences. Some clarity was achieved, which can be presented as follows:

Through the event of 11/9/2003 (a reversal of 9/11/2001), the core of the new Earth space was born. "Earth space" does not refer to physical Earth, but the planet's vital-energetic sphere which makes life on the Earth's surface possible in all its dimensions.

Humanity did not perceive the event consciously enough to be able to incorporate it in any practical fashion. For the powers that work against this epochal change of Earth and humanity (our own alienated powers), the case was quite opposite. They were sufficiently alert to see that this phase was so decisive that without it the change process could not develop further. They have exploited our unclarity to hide the newborn core of etheric space in the most distant corner of the earthly cosmos. In picture language, they have cast the golden ball that is Earth's etheric core into the shaft of a deep well behind our back.

To counter their cunning—if I may continue to use the language of fairy tales—we of the Life Network have called into being a worldwide meditation where the ball that is Earth's center is fetched from the well shaft and afterwards raised into one's own heart center:

- Center yourself in your heart center. From there construct a bridge that leads in a spiral around your head and deep down into the space of your back.
- Make yourself very small and run along the bridge, deep down into the space of your back.
- When you reach the level of your coccyx, you find a deep well there. Let yourself down into the well and seek in its bottom mud for the golden ball.
- After you find it, clean it and take it into your heart. Carry it out of the well shaft, and then back along the same bridge as before into your heart center.
- Let the rays of the golden ball in your heart center stream forth, to connect the newfound center of Earth space with all the various aspects of society, nature and the universe.

It is my opinion, based on this meditation, that one can truly sense that the new Earth space is no longer centered outside of us, in the so-called realm of objectivity. The enormous power of the new space is that its multi-dimensionality is centered in the innermost hearts of many individual people who are prepared lovingly and consciously to nurture the Earth space as their own living and creative space.

In consequence something powerful ensues: The new space has as many centers as there are people who believe in its actualization.

This brings about unbelievable diversity. At the same time, one should not imagine a multitude of "centers," but logically just one single one. That one joins us all, and at the same time we all maintain it.

Sempas, February 12, 2004<br>
Marko Pogačnik

APPENDIX

# Overview of the Exercises

## 1. The Inner Child
These exercises serve to make contact with the various archetypal images of the spiritual and soul planes of humanity.

1.1 To experience the inner child
- Lie down with legs outstretched and be still. Then imagine yourself as a little "child seven days old," lying head down in the water of your belly.
- Suddenly, the process of inversion begins. In the watery region of your lower body the child makes something like a backwards somersault. It turns itself around like a fish in water and stretches upwards, landing in the middle of your upper body.
- The top of its head is to be felt near your throat and its root chakra in the region of your solar plexus. The water has vanished.
- Try to sense the presence of the inner child everywhere. It is the inner Self, called awake, and now focused in your heart-center.

1.2 The divine child—the Higher Self—is handed to you
- You sit for a few minutes sunk deep in inner silence. Then,

when you are ready to hold the divine child in your lap, gently stretch out your arms and ask the Mother of Wholeness, the divine Virgin, to give the child to you.
- You hold the child for a while so as to feel the cosmic quality that streams from within the Christ Child.
- Suddenly, the child in your lap appears to see something interesting on the floor and bends down to touch it—finally bending so far forward that the top of its head points to the Earth's center. Thus it happens that the child turns upside down, as suggested by Jesus' words: What was highest will be lowest, thereby thrusting open the doorway to eternity.
- At this moment you should envision that the child's body turns quickly around as if making a somersault and rises up through your own bodily structure. This motion is the exact opposite of the birth process. You are, so to speak, new born, but this time not through your physical mother's birth canal but by consciously turning round and taking your own path.
- The child is now in the middle of your inner space. You feel the challenge to let go of material forms and projections and instead concentrate on experiencing the qualities of emotions, soul and spirit. How does it feel for the divine core to be awakened within you? It gives you free access to the original space of eternity. What are you going to bring there with you?

1.3 Continuation of the exercise with the divine child
- Imagine that the inner child is fully present and standing in your midsection. Now imagine yourself to be touching various points on the soles of the child's foot with the sensitive fingertips of both your hands. It is recognized that specific zones on the soles of the feet resonate with various physical organs and functions. Relating this to the divine Child, it means that specific points on the soles of its feet resonate with the "organs" and "functions" of the Universal

Wholeness, or, more precisely, with the various powers that ensoul the universe.
- Through the soles of the inner child's feet, connect and anchor yourself to the multi-dimensionality of life.

1.4 The inner maiden as symbol of the soul
- This time, it is Christ as the Father, holding the little girl Mary on his lap, who represents the source of inspiration. Now ask him to hand you the little girl. See how it feels to be holding the archetypal image of your spiritual soul in your lap.
- Then let the child turn upside down and go through the same somersault process as was described above. When she stands up within you, take particular note of the new and blissful quality that is expanding within your being. Is it distinguishable from how the inner child's masculine aspect revealed itself within you, or is it even complementary?

1.5. To speak with the inner child
- Look for a peaceful place and put a cushion in readiness nearby. Let yourself sink into the Silence and center yourself in your midpoint.
- Take your inner child onto your lap and let yourself sense its blessed presence. Then gently stretch out your arms.
- Ask either the divine Virgin or the Heavenly Father to hand you the inner child of a particular person. This person can be one who is already dead.
- Place this person's inner child on the cushion that is ready near you. Then listen to the sound of the conversation taking place between the two children. What feelings or images arise within you from this holy conversation?
- At the end, thank the other person's inner child and hand it back to its divine parents. Your own inner child is restored to your wholeness by the somersault described in Exercise 1.1.

1.6 Help provided by the inner child
Based on the archetype of Anne the Threefold, there is an exercise that can be used to help other people—those who are suffering or have sustained some trauma, or who merely need a loving embrace:
- Imagine you are taking your inner child onto your lap and connecting with it through your feelings. Then, gently stretch out your arms and imagine that your hands are one with the child's tender hands.
- Now you take the next step: Think of the person you want to help and, together with your inner child, take that person's inner child onto both your laps.
- To envision how things are going with the person concerned, you first need to sense that person's presence in your united heart.
- Now the time is ripe for you and your inner child to give the person your joint gift: With your joined hands bless the other's inner child on both the left side and the right. Observe at the same time any changes that can be experienced in the other's inner child.
- After a while, the other person's inner child will wish to say goodbye, and you will hand it back.
- You should then put your hands together in a gesture of thanksgiving.

1.7. The elemental and the divine child within you
This exercise enables us to experience a harmonious relationship between the two partial aspects of our inner Self.
- Stretch yourself out on your back and imagine that two boys are lying very quietly within your body—if you are a woman, they would be two girls.
- The light-skinned boy's scalp reaches as high as your throat. The fontanel of his dark-skinned brother touches your knees,

and the soles of his feet reach your sexual region. There, they rest against the soles of the light-skinned boy's feet.
- You should make sure that you get a very precise feeling for the sensitive contact between the soles of those two pairs of feet. This is the paradisiacal touch that makes possible the birth of our true Self.

1.8. To perceive with the eyes of the soul

To perceive the inner extensions of reality, one should use the eyes of the soul. The phrase "eyes of the soul" puts one in mind of the third eye located behind the forehead. It is certainly true that we are here dealing with the soul's organ of sight. However, the soul does not concentrate its vision on a single point. This is the reason why the third eye, which is really a spiritual organ of perception, is imagined to be a left over from the old patriarchal epoch of human evolution.

The soul represents a fractal—a holographic fragment—of the Goddess. Consequently it displays three aspects, which are:
1. The holistic aspect of the soul (the maiden)—color white.
2. The creative aspect of the soul (the partner)—traditional color red.
3. The transformation aspect (the old wise woman)—color black.

One can further imagine that each of the soul's three faces is provided with an eye. It follows that the soul has not one but three eyes. The three eyes are each assigned to a spot in the physical body that is characteristic of one of the soul's aspects:
1. Behind the middle of the forehead: the eye of the Virgin Goddess within us.
2. At the upper edge of the belly region: the eye of the Partner Goddess within us.
3. At the back, underneath the sacrum: the eye of the old wise woman within us (like the others, it faces forwards).

## 2. The Human Holon

The exercises are put together in such a way that the complete Holon can be constructed stepwise. First comes the centering and rounding-off. Exercises follow for:
- The Holon's vertical axis (the Earth-heaven axis)
- The horizontal left-right axis (Yin-Yang axis)
- The horizontal posterior-anterior axis (dark-light axis)

### 2.1 To find your midpoint

This midpoint may usually be located in the area between the heart and solar plexus chakras, but you may also have a sense of it being situated higher or lower. It is recognized that Eastern cultures tend to center it lower in the belly, while Western ones place it higher, in the heart region. This is the point of inner peace which you should never lose, however dramatic the moment in your life.

- Imagine the energy channel that runs along your backbone to be a transparent tube in which a point of light can move freely up and down.
- Move the point of light up and down for as long as is necessary for you to find the position where your midpoint is located at this moment (your midpoint can, according to conditions, lie in another place).
- Remain for a while centered in your midpoint and enjoy it.

### 2.2. To equilibrate your midpoint

It can happen that you may be quite unable to feel your midpoint, or that it seems to have shifted to one side, somewhere alongside the vertical axis that runs down the spine. In such cases it is advisable to work patiently on centering yourself, so that you can at all times find the place within yourself where you feel "at home."

- Imagine that threads of light are fastened to various points on your lower body. The other ends of these threads are anchored to various points deep in the Earth.
- In addition, imagine that an abundance of threads of light are

fastened to various points on your upper body. The threads end at various points high in the vault of heaven, where they are anchored.
- Now you pull on several of these threads of light—tightening the threads respectively—until you have them perfectly equilibrated at their midpoint.
- You can control the result through Exercise 2.1 detailed above.

### 2.3 To construct the Holon's spherical cloak

The Holon is the autonomous power space of every human being, and is surrounded by various layers of subtle mantles. They represent the personal protective cloak. On one hand they maintain the integrity of the personal space and on the other permit communication with the surrounding ambient field. One can compare them to the fine membranes that close their tiny pores at need.
- Imagine that you are standing within a sphere-shaped space that stretches out on all sides somewhat further than your hand can reach. It is surrounded by several layers of crystal white light forming a protective cloak. The protective cloak extends about 70 centimeters deep into the Earth.
- Make sure that the cloak is well rounded off and tightly closed on all sides. To strengthen it, you can feel or stroke it vigorously from within.

### 2.4 Body cosmogram for rounding off the Holon

This exercise is suited to complete and strengthen the Holon of a place, garden or landscape. It can also be of service to your own Micro-Holon.
- To join yourself to the powers of heaven and Earth, hold both hands in front of your body with one hand pointing to heaven and the other to the Earth.
- Imagine that you are holding a fine membrane

of rainbow colors between your hands. The membrane is so big that it stretches out to the boundary of the Holon that you have chosen for the purpose of this exercise.
- The membrane is not only located above you in the heavenly realm, but reaches down into the Earth beneath you. It follows that the membrane is circular in shape; one half extends into the heights while the other thrusts deep down into the Earth.
- Now, using extreme care, you begin to turn slowly to the right. Take the rainbow membrane with you so that everything above and below the place where you are standing must pass through it. Use the power of your imagination to make sure that the domain you have chosen is in fact guided, centimeter by centimeter to correspond with your turning, through the membrane.
- The exercise is concluded when you have put the membrane through a 360-degree turn. You can repeat the exercise, carrying it out as above but in the opposite direction.

2.5 First Exercise for the Holon's vertical axis (body cosmogram "Star of David")
- Make a triangle with your arms raised above your head. Through this triangle you establish the connection with the cosmic powers above you.
- Slowly lower your arms till they are pointing diagonally downwards. Now let the power of heaven that you have gathered in yourself flow into the Earth.
- Now add to this by bringing your hands together with the elbows thrust outwards to form an earthward pointing triangle. Through this you establish the connection with the power of the Earth.
- Now slowly stretch out and raise your arms till they are pointing diagonally upwards.

Let the power of Earth that you have gathered in yourself flow out into the universe.
- Continue the exercise by again forming a triangle with your arms above your head and connecting yourself with the cosmic powers. The exercise should be repeated with flowing movements at least 10 times.

2.6. Second Exercise for the Holon's vertical axis (The earth-heaven connection)
- Touch the firmament above you with your consciousness. Then, from your upper body project fine threads of light towards the vault of heaven and there clip yourself fast.
- Let yourself feel that you are hanging from heaven with the whole weight of your body. Imagine that your feet are no longer touching the Earth.
- Now you begin to connect yourself with the Earth beneath your feet. From your lower body you send out fine threads of light towards the Earth to anchor them to various points in the Earth's depths. You tighten these threads to the point that you finally touch the Earth.
- You are aware that the weight of your body is being borne in equal parts by Earth and by heaven. Center yourself in your heart and from there hold the scales of heaven and Earth in balance.
- Try to take this quality with you into your rhythms of everyday and not to tread upon the Earth any more heavily then necessary.

2.7 First Exercise for the Holon's Yin-Yang axis (Horizontal left-right)
- Sit or stand upright. Place your right hand on the front of your left shoulder and your left hand on the backward side of your right hip.

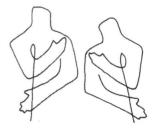

- After a while, change the position of your hands so that your hands and arms describe a smooth arc. Now your left hand travels to the front of your right shoulder and your right hand to the backward side of your left hip.
- Change the position of your hands several times to find your inner balance.

2.8 Second Exercise for the Holon's Yin-Yang axis (The partner within you)
- If you're a man, imagine that a subtle woman also dwells within you. If you are a woman, it's an inner man. Her/his face is turned in the opposite direction to yours.
- Try to give your inner partner as much space as he or she needs: Ask yourself, how does his or her presence feel within me, how am I reacting to it? What can I alter in my life or way of thinking so that the male and the female halves of my Holon are both happy? They should be able to deal with each other in a way that enriches both.

2.9 First Exercise for the horizontal axis from the back to the front side of the body (A hand cosmogram)

The hands are a Holon in themselves and are structured like the human Holon. One can do creative things using the front side of one's hands. In contrast, the backs of the hands appear of little use, but, as is true for the whole body, they are no less important. It is the backs of the hands that guide the delicate movements of the fingers.
- Put your hands in front of your chest, placing them so that the palm of one and the back of the other are facing forward and their edges lie against each other. In this position the hands correspond to the Yin-Yang sign, meaning that the anterior and posterior sides complement each other.
- Then begin to circle the edge of one hand round the edge of the other, so that the edges are constantly touching each

other. The two hands circle round each other like two millstones. The direction is not important.
- After a while, imagine that the "millstones" are not circling in the space in front of you, but within the space of your breast.

The more often this exercise is repeated, the more it will help to pulverize the blockages that separate the anterior space of present-day reality from access to the storehouse of original powers waiting at the back.

2.10 Second Exercise for the horizontal axis from the back to the front side of the body (A body cosmogram)
- While standing, grasp with your hands into your back space and guide them together above your bottom so that the middle fingers touch. From there, connect yourself with the archetypal powers of the Earth's depths.
- Then raise your hands up along the sides of your body until your arms and hands stretch up diagonally out over your larynx. Then bring your hands together so that the middle fingers touch each other again.

- While you are stretching your hands upwards, you should incline your head as far back as possible so as to free up the neck chakra.
- While you are inclining your head backwards and making your arms form a circle in front of your larynx, you should be conscious that the archetypal powers of the back are now changed to the creative powers of the Word.
- Next, open your hands and stretch out your arms so that the powers of the Word can flow out unhindered into the world.

- Following right along, let your arms come down to your sides and bring your head forward in order to repeat the exercise. Practice it several times, one after another, with flowing movements.

2.11 Grounding
- Imagine that you are a cheerful tree. From your lower body strong roots develop, holding you broadly and deeply rooted in the Earth. Using the power of your imagination, follow for a while the path of these roots, which branch out abundantly into the Earth, becoming ever finer and ever more united with the Earth.
- Imagine that from your upper body you simultaneously develop a thick treetop crown whose branches anchor you to various points in the high vault of heaven.

## 3. The Various Extensions of the Human Being
3.1. The experience of the cosmic double
- Seek out a calm and peaceful place to go deep into the inner silence. Imagine that a person similar to you is sitting behind your back—indeed, you are sitting back to back.
- Now, let this cosmic double glide like a breath through your body, so that it appears in front of you. Look it in the eye and at first simply admit its presence into your Holon.
- While you are sitting opposite each other thus, both of you form a rounded space with its center midmost between you. Simply by imagining it, you build this space on the etheric, and on that plane it is real.
- Now, in your imagination, lead both of your figures simultaneously through this middle center. This will cause the etheric space to somersault. Allow the presence of this upside-down space to spread itself out as widely as possible, and get a real sense of its quality.
- Try to incarnate this quality as deeply within you as possible.

3.2 The experience and activation of the lumbar channel
- You may stand, sit or lie down. Find your inner silence and connect with the quality of the cosmos through your crown chakra.

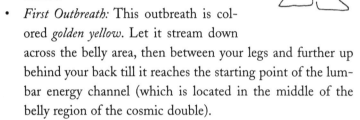

- *First Inbreath:* Draw your first breath in from the middle of the universe; this breath is *white* in color. Draw it down through your body into the solar plexus chakra.
- *First Outbreath:* This outbreath is colored *golden yellow*. Let it stream down across the belly area, then between your legs and further up behind your back till it reaches the starting point of the lumbar energy channel (which is located in the middle of the belly region of the cosmic double).
- *The Second Inbreath* begins at this point. Its flow is now colored *green* to stimulate the regeneration of the lumbar energy channel. Let the flow of the inbreath stream through the channel from its beginning to its outflow point, which is located on the highest part of your physical body between the navel and sex.
- Now you breathe out, and the *Second Outbreath* takes on the color *violet* to effect the transformation of the above-mentioned barriers. Whirling in a myriad of vortices the stream of violet breath passes throughout your Holon on its outward passage.
- *Third Inbreath:* Now the breath is drawn into your heart and there transformed into the perfection of the color *white*. This ensures that the power of the breath that was dispersed throughout the Holon is gathered together again.
- *Third Outbreath:* The assembled impulse is exhaled into the cosmos through the back of the head. There is no color any-

more, but instead a crystal-like clarity that corresponds to the original space of eternity.

### 3.3 The strengthening of the heart channel
- You find your midmost heart and are fully present there.
- From your midmost heart go slowly backwards till you have reached the midmost heart of your cosmic double.
- Imagine a root growth in the form of a mandala that stretches out from the middle of your cosmic double's heart. It contains anchored within it the quality of the first love, whose source is to be found in the midmost heart of your cosmic twin.
- The energy channel resembles the trunk of a tree lying on the ground and grows out horizontally from the middle of the root growth. Its crown is formed in your own midmost heart—a thick crown of leaves and branches that is shot with green throughout.
- After a little while a bud-bearing shoot begins to grow from the middle of the leafy crown. In the region of your breast a wonderful flower unfolds, taking the form of a mandala. Henceforward, its fragrance streams through all the areas of your life.

### 3.4. Joining the three horizontal light channels
This breathing exercise is best carried out while standing.
- Feel yourself anchored in your inner silence and rounded off all through the sphere of your Holon.
- *First Inbreath:* Draw this breath in from the middle of the universe vertically downwards to the midpoint of your lumbar channel (see the left-hand sketch).
- *First Outbreath:* When breathing out,

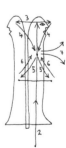

push the breath backwards and forwards through the lumbar channel till it reaches both chakras at the two ends of the channel.
- *Second Inbreath:* Draw this breath simultaneously from the two ends of the lumbar channel diagonally to the middle of your heart.
- *Second Outbreath:* Lead the outflowing stream of your breath diagonally to the two ends of the forehead channel simultaneously.
- *Third Inbreath:* Draw this breath from the chakras at the two ends of the forehead channel back by the same path to the heart center.
- *Third Outbreath:* Breathe the enriched breath out from your heart center into all the breadth of the life surrounding you.
- Repeat the exercise, but draw the first breath from the middle of the Earth and not the universe. Consequently, the order of the breathing sequence is altered: You breathe first through the forehead channel. The lumbar channel follows with the second outbreath (see the right-hand sketch).

3.5. The eight-pointed star

This exercise allows you to get a feeling for the changed shape of your own energy system. The rather clumsy model of the three energy channels layered one over the other has been replaced by the model of an eight-pointed star. Two horizontal rays or points represent the heart channel. The ends of the forehead and lumbar channels are joined to the heart center by four more rays. The vertical axis represents the relationship between Earth and heaven within the personal Holon.

- At the beginning of the exercise, imagine the star with its eight rays as being within your own Holon. The star is positioned at right angles to the anterior surface of your body. Its rays reach almost to the edge of your Holon.
- Now the star begins to turn slowly around towards your back. You should take note of the sensations and feelings that accompany the slow turning of the star and be alert to what develops within you as a result.

**4. The Heart Is the Midpoint**

Because of their key significance, the exercises to support the heart center in its new role have been brought together in their own chapter. The heart represents a Holon in and of itself and, exactly like the principle of the cosmic cross, its vertical and horizontal arms should be integrated in its wholeness.

4.1. Grounding the power of the heart
- Imagine that a tiny golden fragment breaks away from the mandala of your heart center and glides slowly downwards through the watery layers, i.e., the emotional layers, of your belly. Finally it comes to rest on the floor of your pelvis at your pubic bone.
- Prepare yourself to get a precise sense of what happens when that little golden fragment touches the pubic bone. Follow the feelings that will quite possibly explode as they strive to rise upwards. Let your heart bathe in them so as to draw into itself the power of grounded love in all its wholeness.

4.2. Holographic exercise to ground the power of the heart
- Hold both hands below the heart center, horizontal and back to back.
- Imagine that your heart center, in the form of a golden ball, is lying on the surface of the upper hand.
- Slowly lower both hands, still carrying the ball, till they reach

the region of your sex. Let the golden ball stay there for a while and unite yourself with the potential of the elemental power of love.
- Hold the lower hand near your sex while you raise the upper hand till the heart center is back in its place.
- Try to sense the power field of grounded love that has built up between the lower and the upper hands. Take it into your whole being. Breathe with it.
- To further strengthen the heart field through polarization, you can lay the upper hand alternately on the front and back sides of your heart center—the back is located at your back—and perceive the change in relation to your lower hand.

4.3. Third exercise to ground the power of the heart
- Imagine that there is a calm lake lying in your pelvis. Its water is fresh and clear.
- Hanging down from your heart center, fine roots are growing in the direction of the lake. As soon as they reach deep into the lake, they sprout fine shoots of light by which they anchor themselves firmly in the water of the lake.
- What quality of the heart center will this bring into being?

4.4. Transforming the Holon of your head
- You are present in your wholeness, well grounded and centered in the middle of your head.
- You change the physical form of your head into a sphere of light. From now on it will represent the Holon of your head.
- Imagine that you are taking the Holon of your head into your hands and lifting it very carefully off your neck. Very slowly you bring it forwards to the front of your chest and finally place it in your heart space.
- You let the sphere of light that is the Holon of your head rest in your midst until it is completely flooded with the power of your heart.

- Then let that sphere of light rise as gently as a soap bubble until once again it is one with the physical head.

4.5. The power of the heart is joined to the cosmic heart
- Imagine that a wooden gate is standing in front of your midmost heart. The gate has wings on either side and is closed.
- Next you see the two wings slowly opening, and the light of your heart begins to shine.
- The wings open wider and wider until they come together at the back of your heart.
- From the united wings comes a strong root through which your heart center is implanted in the heart of your cosmic twin behind your back. Your heart is open and its beams stream out to every side.

## 5. The Somersault of Space

5.1. A holographic exercise for the somersault

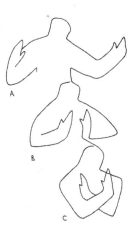

- Stretch your arms out on both sides and imagine that you are touching both sides of the space in front of you (A).
- Now very slowly move your arms horizontally towards each other to the point where they meet. The movement corresponds to the "old" linear space (B).
- Remain in this position for a moment during which the visible aspect of the place is symbolically made null.
- Then continue with the horizontal movement, crossing your arms more and more.
"Soften" your glance and look inward through the "reversed" window formed by your crossed arms. This looking inward is mainly intended to give you a feeling for the reversed quality of space. You can succeed in doing this provided that through

the whole process you make yourself one with your movements and with the space you are watching (C).

5.2. Making yourself familiar with the reversal of space
- *Phase A*: Imagine that the etheric essence of your being (not your etheric body) takes the form of a figure of light and withdraws from your physical body. It moves out through the back of the head to reach the realm of the cosmic double behind your back, making a somersault-like movement so that the etheric essence of your being lands head down in your double's body. Now you have reached the position that can be compared to the first phase of the spatial reversal. The physical being, as always, is still standing normally on its feet. Its etheric essence, however, is hanging head downwards, as if it were hanging down from heaven by its feet. Remain quite calmly in this position for a while.
- *Phase B*: Your being's etheric essence, which is hanging behind your back, now moves into the space of your hips. At the same time, spiral fashion, it folds itself inwards. It begins to turn inwards like a centrifuge, respectively forwards and back. The turning movements purify and at the same time diminish it until there is only a minuscule scrap still remaining in the hollow of your hips. Phase B represents the dramatic reversal in which—as described above—the life force of etheric space is almost lost.
- *Phase C*: Wait patiently. When you feel that the impulse towards a new spatial quality is arising anew from the nothingness, you should let it rise ceremoniously upwards through your body till it reaches the top of your head. Then, spread the new quality out through your entire Holon. Try to incorporate it in yourself as broadly as possible and bring it into your life.

## 6. Clarification and Transformation of Shadow Aspects
6.1. The tear of mercy

First consider where you would
like to guide the tears of mercy and
compassion. It may be for a particular person afflicted with pain, or
for a nature being that dwells in a
place that is suffering severe ecological destruction, or for a country
traumatized by political disturbance.

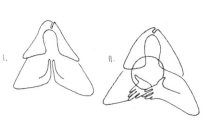

- Fold your hands in front of your breast as if for prayer; your thoughts and feelings are focused on your project. You exchange a few breaths with the divine Virgin, who surrounds you everywhere.
- Open your hands as if opening a book. This movement brings a channel into being. Then ask the divine Virgin—the Goddess of Mercy—to pour a tear into this channel.
- Open your feelings to the subtle dimensions in order to sense when the quality of mercy in the channel begins to vibrate. Then tilt the channel a little downwards so "the tear" starts to move, and guide it in your imagination to the goal that you have projected.
- Keep a strong heart connection with the chosen person, being or place in order to accompany the gift of mercy on its way there. Through the vibrations of your opened heart you should stay present there for a while to trace the effects of the "tear of mercy."
- Give thanks.

6.2. The transformation of the personal shadow aspect
- Ground yourself and visualize your Holon's protective cloak around you. You are connected and centered.
- You are conscious of your cosmic double's presence behind your back. Like a forgotten twin, your double hovers behind you, back to back.
- You check whether he is burdened with the heavy load of one

of your unreclaimed shadow aspects. This may be an aspect already known to you, one with which you are consciously dealing. However, you can also expect to be surprised, and should be open to whatever is ready to change.
- You let the selected shadow of your double glide through your body so completely that every physical cell is touched.
- Then bring the shadow forward till it hovers in front of you and you recognize its "face." Look closely at it, or rather, feel out its presence. Allow its powers to unfold freely before you, however hateful they may prove to be.
- When you sense that these powers have reached their maximum extension, ask for the grace of change.
- After that, bring the shadow, head down and facing forwards, into the vortex of change. There will be a succession of somersaults like the path through a centrifuge, a sequence with many reversals. You can also allow the color violet to flow into the process.
- At the appropriate moment, the concentration on the vortex relaxes somewhat. Now watch what arises from the vortex *at this moment*. It may be a being of light, a light-mandala, a beneficent feeling or much else. The power that was previously frozen in the shadow is now a positive quality and free, and it steps forward.
- Take this power into your heart and let it become part of you. If you have been working to change an estranged fragment of the Earth's soul, the redeemed power will spread out over the world. Give thanks.

6.3. The transformation of the collective shadow aspect
This ritual follows a pattern similar to the one for changing the personal shadow.
- A circle is formed and dedicated to the divine light and the love of the heart.
- An aspect of the collective shadow that has been discussed

and chosen beforehand is then manifested outside the circle, i.e., "behind its back."
- Afterwards, the mega-shadow is drawn through the power ring of the circle to its center, there to be perceived in its true shape—not drawn into the center of the circle through a person's body, which is this case would be too intense an experience. Next the shadow being is delivered over to the vortex of transformation.
- At the end of the process the force, which is now changed and positive, is given back to the appropriate country.

6.4. To work on future events that are in the phase of etheric formation

Bring to mind the archetypal image of the Christ who is holding the little maiden Mary in his lap. He represents the Core of the Universe, and the maiden incarnates the World Soul through whom the Core of the Universe works to protect and nourish creation and impel it forward. The exercise is as follows:

- Center yourself in your midpoint; you are rounded off within the mantle of your Holon.
- Think of the process on whose transformation you would like to work, and envision it as a ball of light hovering within you. The size of the ball will adjust itself of its own accord. Make sure that all the aspects belonging to the relevant process—energetic, feeling- and form-related—are present within the ball.
- Now gently open your hands and ask the Core of the Universe to reach across and give you the little maiden. Let her Being slowly glide through your body until the World Soul is fully present in the realm of your back.
- Next direct your attention to the ball of light once again. Imagine that the presence of the World Soul behind your back is gradually dissolving into a spray of countless drops of water that continually stream through the light-ball to release

the processes of transformation within it. With the help of this penetrating stream the contents of the light-ball will be brought into accord with the harmonies of the Cosmic Whole.
- Now put your hands together and give thanks. This is also the signal for the presence of the World Soul to take its leave.

6.5. A most urgent meditation for peace
- Seat yourself, be still and find the place of peace within yourself. Open your heart.
- Imagine that you are holding in your arms a transparent vessel that is filled with clear water. This vessel filled with water is identical to the power field of the Near East. The field is centered in Jerusalem and extends as far as the Indo-Pakistani border. It embraces the whole of Europe and also the northeastern portion of Africa.
- As soon as the water vessel comes into resonance with the power field of the Near East, its contents can become disturbed or get into other difficulties.
- Water is intelligent. It will show you what to do to make it calm and harmonize any possible stormy effects. Take care that your heart is always touching the water's edge. With the help of your imagination, you can make use of various methods for your purpose:

　　Mix the color blue (peace), violet (change) or green (healing) into the water,
　　Take the vessel into your own heart space and breathe through it rhythmically,
　　Take the vessel into the space of your hips so that it can experience the power of your rootedness,
　　Give the vessel into the hands of your own inner child so that the water can experience the touch of the divine,
　　In addition, you can develop and employ various other possibilities yourself.

- Feeling filled with peace and calm, you hold the water vessel for a while. You sense that the water is in harmony, and the same is happening outside in the world.
- Lastly you spread this quality out over the entire Earth. For this, you imagine that you are holding the Earth between your hands as if it were a ball of water.
- To close the meditation, put your hands together as if in prayer, release yourself from the power field of the Near East, and give thanks.

### 6.6 The inner Big Bang

- Find your midpoint—it is deep in the chalice of your hips, roughly where the lumbar channel is located. Rest there in your mid-point for a while, wholly strong, decided and present. It is worth recommending that you ask for support from your spiritual master or guardian angel, the Archangel Michael or another spiritual helper.
- When the right moment arrives, use your imagination to let a sound like the Big Bang spring out from your midpoint. It is important that this should not be a fiery sound but "cold"—as if a piece of iceberg had broken off and fallen into the polar sea. It is important too that you direct the sound of your Big Bang inwards and not outwards.
- Following the Big Bang, cleansing waves spread like rings outwards from your midpoint. To heighten their purifying power, you can imagine these concentric waves to be bathed in violet light. When they reach the rim of the Holon and are flung back, their color is changed to white in order that they may send the quality of purity through the Holon.
- If necessary, repeat this exercise several more times. Give thanks for the spiritual support that is given you.

### 6.7. Release from your own involvement with the dark powers

- Take yourself into the inner Silence, ground yourself, strengthen your protective sheath and center yourself.
- Be aware of the heart channel that runs horizontally to join the three aspects of your heart chakra: the cosmic heart behind your shoulders, the heart chakra itself and the outflow chakra in front of the breast.
- Build a sphere of light at the end of the heart channel in front of your breast. Decide whether you wish to work on the release of the conflicted soul portion within you, the alienated fractal of a specific place or a darkened fragment of the world soul.
- Gather behind your back the feelings, situations or symbols that relate to the alienated aspect of the soul substance on whose release you will be working. However, you can be open to whatever is standing in line for change.
- Allow the unhappy force that you have collected behind your back to flow through your heart channel into the sphere prepared for it. Accompany its movement with a prayer for its forgiveness and release.
- When the dark soul substance has finished its passage, or wishes it to be broken off, you should close the entry at the rear and completely empty out the channel into the sphere.
- Next you should imagine yourself taking the light sphere into your hands *and uncoupling it from your heart channel.*
- With the help of your imagination, you should now stimulate the process of change within the sphere. You can use colors for the purpose, e.g., violet, or work with a vortex of water, initiate a dance of vibrations within it, etc.
- When the change has got under way, you should call the cosmic messengers to you—the angels—and ask them to take over the light sphere and bear it to wherever the changing of the soul substance contained therein can be victoriously completed. Purified, it shall be given back to the place where it was alienated.
- Give thanks and experience the happiness of liberation.

### 7. The Seven Foundation Stones of the New Ethic
The seven aspects of the new ethical orientation are conveyed by the seven cities of the Apocalypse where there were early Christian communities:

*Ephesus: Love!*
At every moment, let yourself follow the call of your heart. Test yourself whether, in any given situation, you really embody the voice of original love.

*Smyrna: Do not fear!*
Never shy away from whatever your personal or collective fate may be sending you. In every situation preserve your inner peace.

*Pergamos: Change yourself!*
Always be ready to pursue the incessant stream of change. Test out for yourself which aspect of you or your creations is next in line to ask to be changed.

*Thyatira: Be truthful!*
Test yourself to be sure whether in any given moment you are not hiding some aspect of the truth either from yourself or from others. Always investigate your heart and your spirit to ensure that you have not become the victim of self-deception.

*Sardis: Be whole!*
Always remain emotionally aware of your many layered wholeness. Keep the great round of your being embraced in your consciousness and anchored in your mid-point.

*Philadelphia: Be loyal!*
Do not forget who you are and to what ideals you have inwardly pledged your faith. Keep reminding yourself afresh of your spiritual calling.

*Laodicea: Be decisive!*
In every situation you are given various possibilities from which to choose. You are called upon to make decisions. The one thing that you may not do in this epoch of great change is to remain undecided.

## 8. The Changed Elemental Beings

### 8.1 To perceive the "new" elemental beings

To perceive elemental beings, I often use an exercise that starts from the chakra of the earth element, located between the knees. If we humans had a beautiful tail, like a she-wolf's, for example, this chakra would be located at the end of the tail.

- Concentrate on the chakra between the knees and at the same time open yourself to communication with the relevant being.
- To perceive the new quality of the elemental being, guide your focus away from the chakra between your knees and up along your vertical axis. The perceptive point is now located between the belly and the heart.
- Starting from there, open yourself to the presence of the unknown beings.

### 8.2. To experience the beneficial effects of one's personal elemental being

- Center yourself in the midpoint of your heart. Then direct your attention downwards along your spine till you reach the endpoint of the coccyx. Next, galvanize the power concentrated at the base of the spine by rhythmically moving your two hands behind your back. The movement is like that of a dog wagging its tail.
- Intersperse this movement by frequently raising both your hands to grip your head and then, as if they were a comb, drawing them down the length of your spine and then up again. This movement spreads the impulses released by the elemental being through the entire body.

### 8.3. Get in contact with your personal helper

- Go into your inner silence and make sure that your mind has taken up a stand-by position.
- Now touch your earlobes. It is best to rub them lightly for a while between thumb and index finger.

- Now bend your tongue backwards for a moment and open yourself up to the space that has its source behind your back and from there extends outward to all sides—and to the front too. Everything is now ready for a conversation.

### 8.4. Cosmogram for perception of the changed beings of the water element

With this exercise you make the connection by treading the path of resonance.

- Visualize and actually feel that there are fish scales on the sides of your thighs. In addition, imagine that your feet have grown together to form a fish's tail.
- Now draw your fish's tail forward and let it glide slowly upwards, tight against your body.
- The fish's tail slides upwards till its fins come as high as your heart chakra. It should then feel as if your heart's center is being held in the chalice of the fish's tail.
- Now it is possible for you to communicate with the changed beings of the water element.

### 8.5. To perceive the personal elemental being

- Imagine that you are sitting on a mirror. When you look down, you see yourself—your second self. If you are doing the exercise while lying down, imagine that the mirror is leaning against the soles of your feet.
- When you are ready, let the second self—the one you see in the mirror—stand up quickly; you then draw it up to your shoulders.
- Embrace it so that your left hand lies on your heart center and your right hand on your solar plexus.
- How does the presence of the elemental being feel to you? What quality does it bring to the Holon of your being? Caress it for a while; express your thanks or make your request.

8.6. Experience the new shape of your own personality
- Find your inner silence.
- Make contact with your elemental essence while raising your attention slowly and lovingly from the sexual region towards your midmost heart. Somewhere between the solar plexus and the heart's center you will find the point where your elemental aspect has made its home since it joined in the current changes within the elemental world. Rest there in its center for a while.
- Starting from the area of your larynx, you now begin to braid a wreath. It is braided out of several strands of crystal clear water. It runs across the left-hand chakra of the water element down to the middle of your belly, and then across the water element's right-hand chakra back to your larynx.
- After that, direct your attention to the above mentioned center, which is now also the center of the wreath. You are present in the middle of it and you let the wreath around you become round too. You allow its watery substance to transfer itself into the feeling of grounded love that the elemental beings know.
- A watery, spherical space arises in the middle of your Holon. In it are joined the forces of head and mind, as well as the qualities of belly and feeling. Henceforth, whatever you do in everyday life, you will be learning to think, will and act out of this new space of integrated personality.

## 9. Discovery of the Spiritual-Emotional Plane
9.1 Ask his heart
First decide what question you wish to ask, then begin the exercise.
- Imagine that a wooden gate is standing in front of your midmost heart. The gate has wings on either side and is closed.
- Next you see the two wings slowly opening, and the light of your heart begins to shine.

- The wings open wider and wider until they come together at the back of your heart.
- From the united wings comes a strong root through which your heart center is implanted in the heart of your cosmic twin behind your back. Your heart is open and its beams stream out to every side.
- You keep yourself firmly rooted in the heart realm of your cosmic double while you turn to the answer to your question.
- You formulate the answer to the point that you can later make it concrete, and give thanks.

9.2. The animals' personal gift
- Using your imagination, invite the animal you have chosen into your Holon. There you let it swim, chirp or graze—or simply be.
- Ensure that the emotional quality embodied in the animal spreads through all your personal space—to dissolve all the illusions, chaotic circumstances or unclean places that reside in crevices within the sphere of your emotional power. Simply let it happen and follow the process with awe and amazement.
- Then give thanks and let the animal go free.

9.3. The emotional grounding of the soul
- Kneel on the ground.
- You have surely seen plants stretching out almost transparent roots, not into the Earth, but in water. Allow similar transparent roots to grow from your knees into the depths beneath you. As you do so, imagine that the Earth is not composed of a firm and rocky crust, but of water.
- Give every single one of those fine roots a different rainbow color. Let them penetrate the watery depths of the Earth and follow their wave-like streaming.
- Finally, stand up and lift the roots without letting them tear.

The goal of the exercise is accomplished if you succeed in keeping the emotional roots of the soul intact whatever you are doing.

9.4. To change the exaggerated Yang
- *First Inbreath:* Draw the breath out of the width of the universe into the area of your larynx—the color is *crystal white*.
- *First Outbreath:* Breathe out and vertically down to the pelvic floor to build the pillar of Yang—the color is *golden yellow*.
- *Second Inbreath:* Draw the breath from the depths of the Earth into the heart center above—the color is *green*.
- *Second Outbreath:* Breathe the power of Yin from the heart center out into the space in front of you—the color is *blue*.
- To continue the exercise, begin again with the "First Inbreath."

9.5. To uncouple oneself from the old world structure
- Be present in your midmost heart. Sink down into the Silence.
- Imagine that your heart center resembles a golden sphere. This sphere is surrounded by several layers of crystal clear light. These layers are like a fine sieve.
- Using the power of your heart, push these sieve-like layers of light through your energetic and bodily structures, tackling them from all sides. Push them through your emotional fields. All the patterns and forces that belong to the old world are thus sifted from your cells and power fields.
- Ask that all your ties and dependencies that relate to the vanishing old world structure should be withdrawn from your body. In this way, all the "old junk" is brought so far out into the open that you can finally push it from your Holon.
- Then gather the "junk" into a separate ball of energy that is clearly apart from the sphere of your Holon. Give thanks for the teachings that these ties, now severed, have imparted to

you. However, express very clearly that henceforth you will have no more to do with them.
- Then imagine that this ball is guided very quickly out of the remains of the old world into the realm of change that lies in the distance. There it dissolves itself in light.
- Be thankful that you are conscious of the true shape of wholeness.

9.6. Attunement to the new spiritual-emotional fields
- When kneeling, standing or lying down, imagine that you are simultaneously bending backwards and raising your feet up behind you to such a degree that the back of your head touches the soles of your feet. This contact rounds you off, and, at the point where the contact takes place, it makes it possible to connect with the desired spiritual-emotional field.
- Then let the image go, and for a while breathe the desired quality into your body and your Holon respectively. Rejoice for the gift you have been given.

9.7. A way to reinvigorate
If you notice a decrease in the life force in connection with food, dwelling or other particular living situations, you can make use of the following method to stimulate their regeneration:
- Give yourself up to the Silence for a moment. Then, with the help of the imagination, you should lift whatever asks for renewal out of the framework of reality.
- Bring whatever is to be renewed (food, for example) into the realm of one of the three light channels. The heart channel can be recommended in all cases. In certain cases the forehead and lumbar channels may also be appropriate. The lumbar channel is indicated if the treatment has to do with a lack of the source powers of life, the forehead channel if a spiritual blessing is needed. The three channels represent the bridge

that connects the limitless vital powers of the spiritual soul to the reality of every day.
- For example, food: When the meal lies ready before you on the table, briefly shut your eyes and ask for the purification and reinvigoration of the food. In your imagination, take the essence of the meal briefly into your mouth and so bring it into the region where it can be blessed by the forehead channel that runs just above. Then let it sink into your heart space to regenerate it and, finally, still deeper into the pelvis for it to be strengthened with life's original powers through the lumbar channel. Then reunite the essence of the food with the meal on the plate in front of you and give thanks.
- The exercise can be adjusted to the exigencies of the moment and switched around as you wish. Be inventive. It is only important to observe the principle that external vitality is brought about by one's own spiritual-soul powers.

9.8. To center the Earth's core anew
- Center yourself in your heart center. From there construct a bridge that leads in a spiral around your head and deep down into the space of your back.
- Make yourself very small and run along the bridge, deep down into the space of your back.
- When you reach the level of your coccyx, you find a deep well there. Let yourself down into the well and seek in its bottom mud for the golden ball.
- After you find it, clean it and take it into your heart. Carry it out of the well shaft, and then back along the same bridge as before into your heart center.
- Let the rays of the golden ball in your heart center stream forth, to connect the newfound center of Earth space with all the various aspects of society, nature and the universe.

9.9. When nothing helps

Situations may arise when none of the exercises proposed here are of help. Instead, there remains available one unsurpassed recourse: Go on your knees, connect yourself to the spark of light in your midmost heart, and ask. Remain faithful to the Mother, whom we call Earth.

# BIBLIOGRAPHY

**Books in English by Marko Pogačnik**
*A Hidden Pathway through Venice.* Carucci, Rome 1986.
*Nature Spirits and Elemental Beings.* Findhorn Press, Scotland 1997.
*Healing the Heart of the Earth.* Findhorn Press, Scotland 1998.
*Christ Power and the Earth Goddess.* Findhorn Press, Scotland 1999.
*Earth Changes, Human Destiny.* Findhorn Press, Scotland 2000.
*Daughter of Gaia.* Findhorn Press, Scotland 2001.

**Books in German**
*Die Erde heilen.* Diederichs, Munich 1991.
*Die Landschaft der Göttin. Heilungsprojekte in bedrohten Regionen Europas.* Diederichs, Munich 1993.
*Elementarwesen.* Knaur, Munich 1995.
*Schule der Geomantie.* Knaur, Munich 1996.
*Wege der Erdheilung.* Knaur, Munich 1997.
*Geheimnis Venedig. Modell einer vollkommenen Stadt.* Diederichs, Munich 1997.
*Erdsysteme und Christuskraft.* Knaur, Munich 1998.
*Die heilige Landschaft - am Beispiel Istrien,* Hagia Chora, Muehldorf 1999.
*Brasilien, ein Pfad zum Paradies.* Edition Ecorna, Ottersberg 2000.
*Die Erde wandelt sich.* Knaur, Munich 2001.
*Die Tochter der Erde.* AT Verlag, Aarau 2002.
*Erdwandlung als persönliche Herausforderung.* Knaur, Munich 2003.
For translations in other languages see the book list at the Internet address below.
Also mentioned: Ajra Miska, *Von der Ewigkeit berührt.* Edition Ecorna, Ottersberg 2001.

*For Further Reference*
More exercises and information on the ongoing processes of Earth Change can be found in my Internet site under "Earth Changes"—www.ljudmila.org/pogacnik/ or www.pogacnikmarko.org—or in the column on Earth Change that I regularly write in German for the geomantic magazine *Hagia Chora* (www.geomantie.net).